ALGEBRA EXAMPLES

POLYNOMIAL FACTORIZATIONS 1

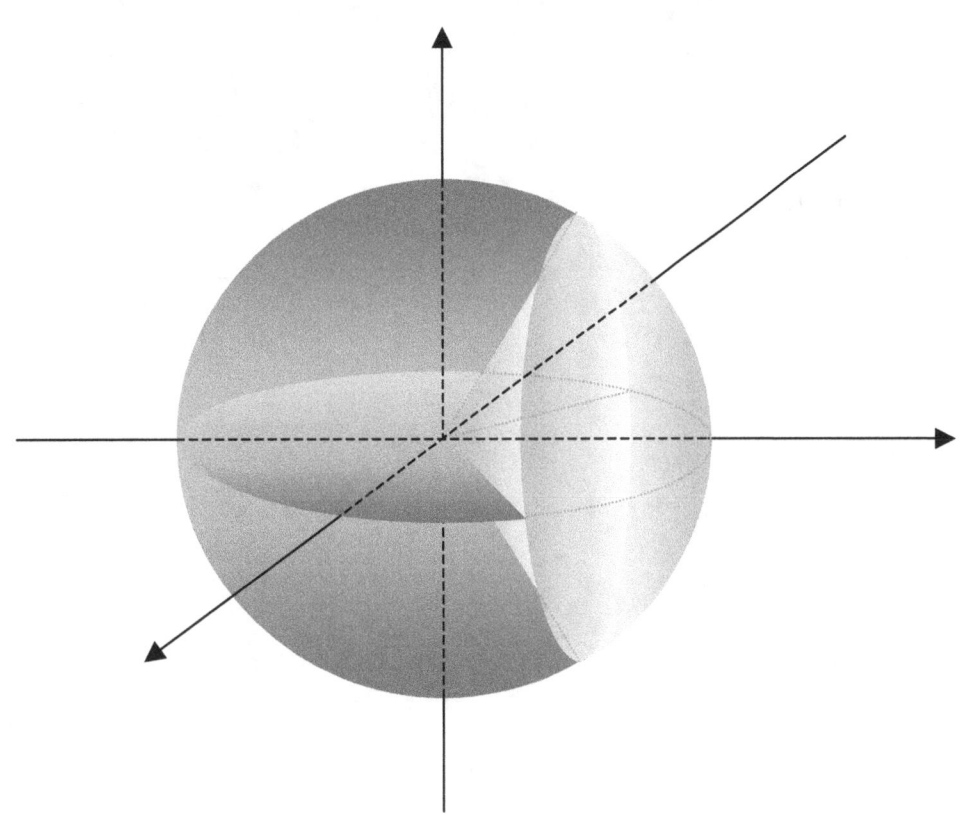

Seong R. KIM

Dear students:

Students need the best teacher, so you need examples, because examples are the best teacher. All the examples here are fully worked, and explain **how** the basic and essential tools in math are made, together with **what** they are, **how** they work, and **how** to work with them. Such tools include numbers, formulas, identities, equations, laws, etc.

Examples here begin with easy ones, of course. Covering every meter and yard properly, we can cover thousands of miles and kilometers. And it is particularly the case in math.

Of those examples therefore, some might even look too easy for you. It's not that easy though, to come up with those examples. Anyways, the bigger and the taller the tree, the deeper and the stronger the root.

Doing math, we work with ideas and run ideas, because every thing in math is an idea. A number is an idea, for instance, and the same is true for a line or circle, too. And putting ideas together, we build another, which becomes the base or an element of another, and each is connected. And that's the way your math grows. So you get to build a circuit, and sometimes, need to fill the gap or repair the circuit so that you get the sense of it.

So your calculation runs properly, and you get the problem solved.

The examples have been made and arranged so that they get tougher (or sometimes easier for some reason) as you proceed with them. In particular, similar examples with some variations are strategically repeated so that you can get the ideas or the tools tricky or complicated, and can get them mastered.

This book is however, nothing but a bunch of examples until you get it powered. How then, to get it powered, and make it run and work for you?

Just read it, and then, do each example in writing. And it is important to note that you do it in **your** writing. Just watching someone doing it, you just only feel that you can do it. If you do it, you can do it, but if you don't, we can hardly. It's a cliché, of course, but is always true that knowing is one thing and doing is another.

I've been helping students grow, take care of, and run their own math. The area covers algebra and geometry for high school or college students, and is especially for equations (for unknowns or curves), functions, and their graphs, which are the basic elements in calculus, which's been the core of my interest from my early age in high school.

Of my students, some are quite poor in math, and thus, are afraid of or hate math, some require special education because of exceptional intelligence, some are smart enough, some are naïve and diligent, some are clever but lazy, and most behave in general. All the students are badly after though, one thing in common: a strong and secure math skill. It is of course, the prime objective of my work, and I'm always happy to and eager to help them achieve it. The problem was however, that many of them wanted it to be purchased. And the question is, can we buy it?

We can buy the means, of course. And a solid math skill is feasible, too. We know however, we can't buy love, and the same is true for the math skill, too. It's not what we can buy or sell, and not what we can give or take. It is however, what we can grow, and need to grow. Your math grows as much as you grow and take care of it. So does mine.

What math then, do students most often do or use in high schools or colleges?

It is algebra and geometry. What algebra though?

Elementary algebra, of course
Doing the algebra, we work with numbers (many in kinds), constants, variables, ratios, rates, expressions, equations, inequalities, functions, identities, formulas, laws, etc., together with signs and symbols. And if we want to do algebra properly, we want to know their natures and how they mingle with each other.

So studying math ideas or tools, you want to know **what** they are, **how** they work, and **how** to work with them or **what** to do with them.   What then, about the geometry?

Basically, the geometry has much to do with shapes, positions, and angles. The shapes begin with triangles and circles, and move on to rectangles, squares, parallelograms or rhombuses, trapezoids, tetragons, other polygons, polyhedrons, etc.

Doing the geometry, too, though, we need to do the algebra stated above. So it is analytic geometry, often called coordinate geometry, too. And doing it, we can specify positions using coordinates. So in the geometry, basically, we work with graphs. Putting a math idea in a graph, we can not only effectively think about it but actually see it, too, and therefore, can efficiently work with it. What idea then, is it?

The idea begins with a point, line, parabola, circle, ellipse, and hyperbola, called a conic section or basic curve, and then, moves on to other curves, planes, surfaces, volumes, and other objects in various dimensional spaces, together with vectors.

And using an angle, we can specify an amount of turn or change in direction.

So learning, using, or applying those ideas or math tools, we get to solve problems.

And this book can help. It can help learn them, and use them so that you can navigate to find solutions to problems. And in particular, it can help come up with answers to those **what**s and **how**s stated above. So it can help you grow and run your own math, and thus, can help achieve your solid math skill.

It is however, not a magic book giving you a math skill of high caliber overnight. And it can have many mistakes, too. There is no magic, and math is full of facts and ideas. And it is after all, not me and not your teacher but you who put together some of those facts and ideas, and understand it. Putting facts and ideas together, understanding it, and taking care of what you have learned, you grow your math. And this book can help.

This is a book of examples designed to help you grow your math, and assumes that you are a real beginner. This book requires though, time and effort, the amount of which need to be substantial, too, but will be worth it. That's because you want a substantial achievement, and will get it. And probably, you will get to see this book helping you get there much faster than expected. And then, you will get to see the way math runs.

In math, everything is an idea. So is a problem. And solving it, we put it many different ways. For instance, while expanding or reducing it, or modifying or converting it, we keep searching for the solution, approaching the solution, and eventually, can get there. So don't look for the solution outside the problem. The solution is inside the problem if the problem is properly made.

If it is not, no solution is the solution. And in fact, it is often the case a problem itself is the solution. We can put a problem in many different ways, and eventually, can end up with the solution.   How come then, is the solution no other than the problem?

For instance, the solution to $3232 \div 101$ is 32.   And we can put it this way:

$$3232 \div 101 = \frac{3232}{101} = \frac{32 \times 101}{101} = \frac{32}{1} = 32 \implies 3232 \div 101 = 32.$$

And we can get this, too: $32 \implies 3232 \div 101$.   How?

$$32 = \frac{32}{1} = \frac{32 \times 101}{101} = \frac{3232}{101} = 3232/101 = 3232 \div 101.   \text{Too easy?}$$

For another instance, the solution to $ax^2 + bx + c = 0$ is: $x = \frac{-b \pm \sqrt{b^2 - 4ac}}{2a}$, which is called the quadratic formula.   How come then, is the solution no other than the problem?

We can put it this way:

$$x = \frac{-b \pm \sqrt{b^2-4ac}}{2a} \implies 2ax = -b \pm \sqrt{b^2 - 4ac} \implies 2ax + b = \pm \sqrt{b^2 - 4ac}$$

$$\implies (2ax + b)^2 = b^2 - 4ac \implies 4a^2x^2 + 4abx + b^2 = b^2 - 4ac$$

$$\implies 4a^2x^2 + 4abx = -4ac \implies ax^2 + bx = -c \implies ax^2 + bx + c = 0.$$

And we can get this, too: $ax^2 + bx + c = 0 \implies x = \frac{-b \pm \sqrt{b^2-4ac}}{2a}$.   How?

$$ax^2 + bx + c = a(x^2 + \tfrac{b}{a}x) + c = a(x^2 + \tfrac{b}{a}x + \tfrac{b^2}{4a^2} - \tfrac{b^2}{4a^2}) + c = a(x^2 + \tfrac{b}{a}x + \tfrac{b^2}{4a^2}) - \tfrac{b^2}{4a} + c$$

$$= a(x + \tfrac{b}{2a})^2 - \tfrac{b^2-4ac}{4a} = 0 \implies a(x + \tfrac{b}{2a})^2 = \tfrac{b^2-4ac}{4a} \implies (x + \tfrac{b}{2a})^2 = \tfrac{b^2-4ac}{4a^2} \implies x + \tfrac{b}{2a} = \pm \sqrt{\tfrac{b^2-4ac}{4a^2}}$$

$$\implies x = -\tfrac{b}{2a} \pm \tfrac{\sqrt{b^2-4ac}}{2a} = \tfrac{-b \pm \sqrt{b^2-4ac}}{2a} \implies x = \tfrac{-b \pm \sqrt{b^2-4ac}}{2a}.$$

And we call the set of processes above, algebra.

So if a problem is well defined, that is, if it makes sense, we should be able to get it solved the way below:

A problem $\Rightarrow$ ... $\Rightarrow$ ... $\Rightarrow$ the solution, and thus: **the problem $\Rightarrow$ the solution**.

So solving a problem, we put it many different ways so that we can get to the solution.

And that's the way, math runs.

May your math run very well.

Seong R. Kim

B.S. Math. Michigan Tech. Univ.   M.S. Math. Rensselaer Polytechnic Institute

Notes:

This book is about factorizations of polynomials.
Factorization is often called factoring, and is a math operation.

Factorizing a polynomial, that is, factoring a polynomial, we get to at the same time do all the four arithmetic operations: additions, subtractions, multiplications, and divisions. So we get to do a lot of mental math doing factorizations. Doing them, we don't just do mental math but we also need to apply rules or laws between the operations above, and the operations are done on not just numbers but expressions, too. So we need to do a lot of algebra.

Factorizing a polynomial, we break it into pieces or parts, and put them together to get another expression, which is a product of expressions as $2x(x + 1)(x - y + 3)$. And doing algebra, we often need to do such math operations to change expressions. Why changing them?

That's simply because we want to get to the solutions.
What actually, that is, physically, connects problems to solutions is algebra.
With algebra skill, together with your creativity, you can actually solve problems.

And most expressions are made of polynomials. So if you want to build strong skill in algebra, you want to be good at factorizing polynomials as well as integers.

And all the basics and many ideas on polynomial factorization are covered in two books as well as in one book. And the two books are as follows:

ALGEBRA EXAMPLES POLYNOMIAL FACTORIZATIONS 1

ALGEBRA EXAMPLES POLYNOMIAL FACTORIZATIONS 2

And the two books are combined into one book, which is below:

ALGEBRA EXAMPLES POLYNOMIAL FACTORIZATIONS

So either way, the books explain how to manipulate polynomials, that is, how to change or alter, convert, or modify expressions so that you can come up with the ones you need. The ones are solutions, of course. And that's what polynomial factorizations are about.

The books do not just explain. But they help you, too, follow steps to the solutions so that you can see how you can change expressions, and see how calculations can flow.

With strong foundation of algebra, you can do a lot, and of course, can do problems very well, too. And following steps in factorization processes, that is, changing expressions or taking alternatives shown in the books, you can grow much power in algebra.

So the books will get you not only polynomial factorizations but enhancement of your algebra, too. You will thus, soon be able to change or alter, convert, or modify math expressions so that you can get to the solutions fast.

Contents

In POLYNOMIAL FACTORIZATIONS 1

The Preview of the Contents

In POLYNOMIAL FACTORIZATIONS 2

$$(x + y)^2 = x^2 + 2xy + y^2.$$

$$(x + y)^3 = x^3 + 3x^2y + 3xy^2 + y^3.$$

$$(x + y)(x - y) = x^2 - y^2.$$

$$(x + y)(x^2 - xy + y^2) = x^3 + y^3.$$

$$(x^2 + xy + y^2)(x^2 - xy + y^2) = x^4 + x^2y^2 + y^4.$$

$$(x + a)(x + b) = x^2 + (a + b)x + ab.$$

$$(ax + b)(cx + d) = acx^2 + (ad + bc)x + bd.$$

$$(x + a)(x + b)(x + c) = x^3 + (a + b + c)x^2 + (ac + bc + ca)x + abc.$$

$$(a + b + c)^2 = a^2 + b^2 + c^2 + 2(ab + bc + ca).$$

$$(a + b + c)(a^2 + b^2 + c^2 - ab - bc - ca) = a^3 + b^3 + c^3 - 3abc.$$

Suppose both a and $b \neq 0$, and both m and n are integers. Then, we get:

0. $a^m a^n = a^{m+n}$ **1.** $a^m / a^n = \dfrac{a^m}{a^n} = a^{m-n}$ **2.** $(a^m)^n = a^{mn}$

3. $(ab)^n = a^n b^n$ **4.** $(a/b)^n = \left(\dfrac{a}{b}\right)^n = a^n / b^n = \dfrac{a^n}{b^n}$

Suppose both a and $b > 0$, and m and n both are integers nonzero. Then, we get:

0.1. $a^{\frac{1}{n}} b^{\frac{1}{n}} = (ab)^{\frac{1}{n}}$. **1.1.** $\dfrac{a^{\frac{1}{n}}}{b^{\frac{1}{n}}} = \left(\dfrac{a}{b}\right)^{\frac{1}{n}}$. **2.1.** $(a^{\frac{1}{n}})^m = (a^m)^{\frac{1}{n}}$.

3.1. $(a^{\frac{1}{n}})^{\frac{1}{m}} = a^{\frac{1}{mn}} = (a^{\frac{1}{m}})^{\frac{1}{n}}$. **3.2.** $(a^{mp})^{\frac{1}{np}} = (a^m)^{\frac{1}{n}}$, where p is a nonzero integer.

1. Suppose M, N, and $b > 0$, but $b \neq 1$, and we have: $A = \log_b M$, and $B = \log_b N$.
Then, we get: $A - B = \log_b M - \log_b N = \log_b \frac{M}{N}$.

2. Suppose that M and $b > 0$, but $b \neq 1$, and that we have: $E = \log_b M$.
Then, we get: $PE = P \log_b M = \log_b M^P$.

3. Suppose that a, b, C, and $D > 0$, but a and $b \neq 1$, and that we have: $\log_a C = \log_b D$.
Then, we get: $\log_a C = \log_b D = \log_{ab} CD$.

4. Suppose that a, b, C, and $D > 0$, but a and $b \neq 1$, and that we have: $\log_a C = \log_b D$.
Then, we get: $\log_a C = \log_b D = \log_{\frac{a}{b}} \frac{C}{D} = \log_{\frac{b}{a}} \frac{D}{C}$.

5. $\log_b b = 1$, and $\log_b 1 = 0$. **6.** $\log_b A = \dfrac{\log_c A}{\log_c b}$.

7. $\log_b A = \dfrac{1}{\log_A b}$.

Note:

The drawings or graphs in this book are not exact, and are approximate or conceptual ones.

$\in$	"$a \in B$" means that a belongs to B. "$p, q,$ **and** $r \in W$" means that $p, q,$ and r belong to W.
$\Rightarrow$	"$A \Rightarrow B$." means that A implies B.
$\equiv$	$A \equiv B$ means that A and B are identical to each other.
$\neq$	$A \neq B$ means that A is not equal to B.
$|A|$	The magnitude of A. For instance, $|\text{-}1| = |1| = 1$.
$\therefore$	Therefore
$\Leftrightarrow$	"$A \Leftrightarrow B$" means "If A then B." and "If B then A." We can read $A \Leftrightarrow B$ as "A if and only if B." In such a case, we can say that $A = B$.
Δx and Δy	Suppose that (x_1, y_1) and (x_2, y_2) are two points in the x-y plane. Then, we get either of the two below. $\Delta x = x_2 - x_1$, and $\Delta y = y_2 - y_1$. $\Delta x = x_1 - x_2$, and $\Delta y = y_1 - y_2$.

Distance Formula

Suppose that d is the distance between two points (x_1, y_1) and (x_2, y_2) in the x-y plane. Then, we get: $d^2 = (\Delta x)^2 + (\Delta y)^2$.

₀.Monomials & Polynomials

Doing math, we use a mathematical expression, often called briefly, a math expression. And more briefly, we usually just call it an expression.

And of those expressions, the most often used are monomials and polynomials, together with numbers, of course. Numbers are math expressions showing values.
What then, is a monomial?

Calling briefly a monomial expression, we just call it a monomial.

And the simplest is a power of a variable, where the exponent is a positive integer. So if for instance, x is a variable, x itself can be a monomial, because $x = x^1$, which is a power of x and the exponent is 1. And a monomial can be x^2, x^3, x^4, or such.

And since the exponent is a positive integer, we do *not* take as monomials such expressions as follows: $\sqrt{x} = x^{\frac{1}{2}}, x\sqrt{x} = x^{\frac{3}{2}}, \frac{1}{x} = x^{-1}, \frac{1}{\sqrt{x}} = x^{-\frac{1}{2}}, \frac{1}{x^2\sqrt{x}} = x^{-\frac{5}{2}}$, etc.

And a monomial can be a product of powers, each of which is a power of a variable where the exponent is a positive integer.
So assuming for instance, x and y are variables, we can take xy as a monomial, because $x = x^1$, and $y = y^1$. And we can call monomials xy^2, x^3y, x^2y^4, and such.

Also, taking a product of a number and a monomial, we get another monomial.

And taking a product of a constant and a monomial, we get another monomial, too.

So assuming for instance, x and y are variables, and **a and b are constants**, we can make monomials as follows:

$-x$, $-y^2$, $2x$, ax, $-by$, aby, $3ax$, $3abxy$, ab^2xy^3, $-x/2$, $\frac{xy}{2a}$, $-\frac{2x^2y^3}{3ab}$, $\frac{2x^2y^3}{3a^2b}$, $5xy\sqrt{a}$, $axy\sqrt{2}$, etc.

(If not sure of constants and variables, refer to **ALGBRA EXAMPLES BASIC FUNCTIONS**.)

• Next, adding together two monomials, or adding a monomial to a number or a constant, we get an expression called a *binomial* expression, often just called briefly a binomial.

And note that subtracting an object means adding the negative of the object.

So for instance, we get: $x - y = x + (\text{-}y)$, which is therefore, a binomial, too.

Making thus, some more binomials, we can get:

$x+y$, $x-1$, $2x+y$, $2xy-ay$, x^2-2, y^2-xy, xy^2+a, $2axy^2-3ab$,   etc.

What then, about: $\frac{1+x}{b}$, $\frac{x+y}{a}$, $\frac{ab+x}{2}$, and $\frac{ab+xy}{2}$?

They are binomials, too, if a and b are constants, of course.

That's because we have: $\frac{1+x}{b} = \frac{1}{b} + \frac{x}{b}$, $\frac{x+y}{a} = \frac{x}{a} + \frac{y}{a}$, $\frac{ab+x}{2} = \frac{ab}{2} + \frac{x}{2}$, and $\frac{ab+xy}{2} = \frac{ab}{2} + \frac{xy}{2}$.

Note however, we do not take as binomials $x+\frac{1}{x}$, $\frac{1+x}{xy}$, $\frac{x+1}{ay}$, $\frac{ab+x}{2y}$, $\sqrt{x}+1$, etc. Why not?

That's because every exponent applied to each variable has to be a positive integer, and we have: $\frac{1}{x} = x^{-1}$, $\frac{1+xy}{xy} = \frac{1}{xy} + \frac{xy}{xy} = \frac{1}{xy} + 1 = (xy)^{-1} + 1 = x^{-1}y^{-1} + 1$, and $\sqrt{x} = x^{\frac{1}{2}} = x^{0.5}$.

• And next, if adding together three monomials, or if adding to a binomial a monomial, a constant, or a number, we get an expression called a *trinomial* expression, often just called briefly a trinomial. So for instance, making some trinomials, we can get:

$$x + 2x + y, \quad x - 2xy + 3y, \quad 2x + y - 1, \quad 2xy - ay + 1, \quad x^2 - 2y^2 + ay, \quad y^2 - 1 + x^2,$$

$$xy^2 + ab + 1, \quad 2axy^2 - 3ab + 2x^2, \quad x^2 + x + 1, \quad x^3 - 2x + 5, \quad \text{etc.}$$

And as in the case of a binomial, in all trinomials, every exponent applied to each variable has to be an integer positive, too.

So for instance, $x + y + \frac{1}{x}$ and $\frac{1 + 2x + y}{xy}$ are not trinomials.

• What then, is a polynomial?

Calling briefly a polynomial expression, we just call it a polynomial.

And in fact, 'poly' means 'two or more', so a polynomial means a sum of two or more monomials. So adding together monomials, we get a polynomial.

Also, adding monomials to constants, or numbers, we get a polynomial, too.
So a binomial or a trinomial can be called a polynomial, too.
And thus, the smallest polynomial is a binomial.

So for instance, the expressions below are polynomials:

$$x + 2, \quad x + y, \quad x + y + 1, \quad x - 2xy + 3, \quad 2x^2 + 3xy - x + y, \quad x^3y - ay + 3x^2y - 2x + 4,$$

$$x^2 - 2y^2 + bxy + xy^2 - a + 3, \quad 2axy^2 - 3aby + bx^2 - 4, \quad x^5 + 2x^4 + x^3 - 3x + 5, \quad \text{etc.}$$

And for another instance, assuming x and y are variables, and a and b are constants, and setting: $P = x^3 + 2x^2y - a(b - 2x) + \frac{x}{2b} + \frac{3(y+b)}{4} + \frac{a+1}{2}(x^3 + y) + 7 + b$, we can say that P is a polynomial.

And simplifying the polynomial P above, we can put it the way below, too:

$$P = x^3 + 2x^2y - ab + 2ax + \frac{x}{2b} + \frac{3y}{4} + \frac{3b}{4} + \frac{a+1}{2}x^3 + \frac{a+1}{2}y + 7 + b.$$

And as in the case of binomials or trinomials, in all polynomials, every exponent applied to each variable has to be an integer positive, too.

So for instance, $x - y + 1 + \frac{1}{x}$ and $\frac{1+2x+y+y^2}{xy}$ are not polynomials.

And also, we can say that a polynomial is a sum of terms.
• What is a term though?

In a polynomial, a term is a number, a constant, or a monomial.

So the polynomial P below is made of 11 terms,

$$P = x^3 + 2x^2y - ab + 2ax + \frac{x}{2b} + \frac{3y}{4} + \frac{3b}{4} + \frac{a+1}{2}x^3 + \frac{a+1}{2}y + 7 + b.$$

And the two terms 7 and b in the polynomial P are called *constant* terms.

Why is 7 a constant term though?

Not only a constant but a number, too, is constant, because it doesn't change its value.

So7 and b are constant terms. And in particular, 7 is called a *numeric* term, too.

And if terms are monomials where the variables are the same, and the same exponents are applied to the variables, we call those terms *like*-terms.   So for instance:

2x, **-3x**, and **5x** are like terms.

2ax, **-5ax**, and **3ax** are like terms.

3ax², 2ax², 5ax², -3ax², and $\frac{2}{3}ax^2$ are like terms.

$\frac{2}{xy}$, $\frac{5}{xy}$, and $\frac{3}{2xy}$, that is, **2x⁻¹y⁻¹, 5x⁻¹y⁻¹**, $\frac{3}{2}x^{-1}y^{-1}$ are like-terms, not monomials though.

$\frac{2a}{xy}$, $\frac{5a}{xy}$, and $\frac{3a}{2xy}$ are like-terms, but are not monomials.

And $\frac{2ab}{x^2y}$, $-\frac{5ab}{x^2y}$, $\frac{3ab}{2x^2y}$, and $\frac{\sqrt{3}ab}{x^2y}$ are like-terms, too, but of course, are not monomials.

So some like terms are monomials, and some are not.

Is there then, anything good about like terms?

We can add them together, and make expressions simpler. So for instance, we can get:

2x + (-3x) + 5x = 2x − 3x + 5x = (2 − 3 + 5)x = 4x.

2ax + (-5ax) + 3ax = (2a − 5a + 3a)x = 0·x = 0.

$$\frac{2ab}{x^2y} + (-\frac{5ab}{x^2y}) + \frac{3ab}{2x^2y} + \frac{\sqrt{3}ab}{x^2y} = \frac{2ab-5ab+3ab+\sqrt{3}ab}{x^2y} = \frac{(2-5+3+\sqrt{3})ab}{x^2y} = \frac{\sqrt{3}ab}{x^2y}.$$

1.0. Polynomial Arithmetic 1

What is polynomial arithmetic?

Normally, doing arithmetic, we do it with numbers as integers, fractions, or decimals, etc. And doing it, we do operations of four kinds: additions, subtractions, multiplications, and divisions.

So doing arithmetic with numbers, we say we do *number arithmetic*, and doing number arithmetic, we add numbers together, subtract a number from another, multiply a number by another, or divide a number by another.

And the same is true for polynomials, too.
So we can do such four operations with polynomials, too.

And thus, doing arithmetic with polynomials, we do polynomial arithmetic.
What then, is the difference between the two kinds in arithmetic?

In number arithmetic, we use numbers, and then, the operations produce numbers only. Doing polynomial arithmetic though, we can use numbers and others, too, along with polynomials, and the others can be constants or monomials. So for instance, we can add to a polynomial, numbers, constants, or monomials as well as polynomials.

On top of that, we do not always get polynomials as results.
That's because we can get as results not only polynomials but numbers, monomials, or even expressions that are not polynomials, too.

So let's have a close look at now, what we can get from polynomial arithmetic.

First, even if doing an addition or multiplication, we don't always get a polynomial. Why not always, though?

Doing additions, we can get 0, which is a number, and not a polynomial.

For instance, adding together $x - 2y$ and $2y - x$, we get:
$x - 2y + (2y - x) = x - 2y + 2y - x = 0$.

Taking normal examples however, we can have cases as follows.

Adding a number 1 to a polynomial $3x + y$, we get: $3x + y + 1$, which is a polynomial.

Adding a monomial $2x$ to $y + x^2$, we get: $y + x^2 + 2x$, which is a polynomial.

Adding a constant c to $x + 2x^2$, we get: $x + 2x^2 + c$, which is a polynomial.

Adding a polynomial $x + 1$ to $x + 2x^2$, we get: $2x + 2x^2 + 1$, which is a polynomial.

Next, doing multiplications, we can get 0, too.

Multiplying in fact, anything by 0, we get 0.

For instance, we get: $0 \cdot 2 = 0$, $-4 \cdot 0 = 0$, $x \cdot 0 = 0$, $0 \cdot y = 0$, $0 \cdot (2x^2 + 4y + 7xy^3 + 5) = 0$, etc.

Taking normal examples though, we can have cases as below:

Multiplying **2** by $3x + y$, we get: $2(3x + y) = 6x + 2y$, which is a polynomial.

Multiplying $2x$ by $y + x^2$, we get: $2x(y + x^2) = 2xy + 2x^3$, which is a polynomial.

Multiplying c by $x + 2x^2$, we get: $c(x + 2x^2) = cx + 2cx^2$, which is a polynomial.

Multiplying $x + 1$ by $x + 2x^2$, we get:

$(x + 1)(x + 2x^2) = x^2 + x + 2x^3 + 2x^2 = 2x^3 + 3x^2 + x$, which is a polynomial.

Next, doing subtractions or divisions, we *often* get results that are not polynomials.

Subtracting 1 from $3x + y$, we get: $3x + y - 1$, which is a polynomial, of course.

Subtracting however, $3x + y$ from $3x + y + 1$, we get **1**, not a polynomial but a number.

Subtracting $3x + y$ from $3x + y$, we get 0, which is a number, too.

Subtracting $3x$ from $3x + y$, we get y, which is not a polynomial but a monomial.

Subtracting $3x + y + c$ where c is constant, from $3x + y$, we get c, which is a constant.

And moving next, on to divisions, we can have cases as follows:

Dividing $3x + y$ by **2**, we get: $\dfrac{3x + y}{2} = \frac{3}{2}x + \frac{1}{2}y$, which is a polynomial, of course.

Dividing however, $6x + 2y$ by $3x + y$, we get: $\dfrac{6x + 2y}{3x + y} = \dfrac{2(3x + y)}{3x + y} = 2$, which is not a

polynomial but a number.

Dividing $x + x^2$ by $x + 1$, we get: $\dfrac{x^2 + x}{x + 1} = \dfrac{x(x + 1)}{x + 1} = x$, which is a monomial.

Dividing $cx^2 + cy$ where c is constant, by $x^2 + y$, we get:

$$\frac{cx^2 + cy}{x^2 + y} = \frac{c(x^2 + y)}{x^2 + y} = c, \text{ which is a constant.}$$

Dividing $x^2 + y$ by x^3, we get: $\dfrac{x^2 + y}{x^3} = \dfrac{x^2}{x^3} + \dfrac{y}{x^3} = \dfrac{1}{x} + \dfrac{y}{x^3}$, which is not a polynomial.

Between the two kinds in arithmetic though, there is not only a difference, but the same, too.   What then, is the same?

There are several laws, shared by both, polynomial arithmetic and number arithmetic. And we call them basic laws in arithmetic.

So the basic laws that apply in number arithmetic apply in polynomial arithmetic, too.

That is to say that the fundamentals are the same in both arithmetic.
And they will be covered in the next section.

1.1. Polynomial Arithmetic 2

There are several laws that polynomial arithmetic and number arithmetic share.
And we call them basic laws in arithmetic.
So the basic laws that apply in number arithmetic apply in polynomial arithmetic, too.

That is to say that the fundamentals are the same in both arithmetic. And they are probably well-known to you. Let's now however, go over the arithmetic fundamentals, simply because they are so important.   *The more basic, the more important.*

To begin with, we have a basic law as below:

• If we multiply (or divide) by the same both the numerator and the denominator, in a fraction, of course, the value of the fraction does not change.

So suppose for instance, in operations below, *A*, *B*, and *C* can be numbers, constants, monomials, or polynomials, and can be expressions in other kinds, too.

Then, assuming *B* and *C* $\neq$ **0**, we can get:

$A \div B = (AC) \div (BC)$. In other words, $\dfrac{A}{B} = \dfrac{AC}{BC}$.

$A \div B = (A \div C) \div (B \div C)$. That is, $\dfrac{A}{B} = \dfrac{\frac{A}{C}}{\frac{B}{C}}$. (It's in fact, because of the law above.)

In sum, we can have: $A \div B = (AC) \div (BC) = (A \div C) \div (B \div C)$, that is, $\dfrac{A}{B} = \dfrac{AC}{BC} = \dfrac{\frac{A}{C}}{\frac{B}{C}}$.

Note that *B* and *C* above cannot be 0, simply because no denominator can be 0, which is in fact, another fundamental law in arithmetic. So keep in mind no division by 0 exists.

And taking some instances with numbers, we can have:

$15 \div 30 = (15 \cdot 3) \div (30 \cdot 3)$. In other words, $\dfrac{15}{30} = \dfrac{15 \cdot 3}{30 \cdot 3}$.

$15 \div 30 = (15 \div 3) \div (30 \div 3)$. That is, $\dfrac{15}{30} = \dfrac{\frac{15}{3}}{\frac{30}{3}}$. It's simply because we get: $\dfrac{\frac{15}{3} \cdot 3}{\frac{30}{3} \cdot 3} = \dfrac{15}{30}$.

In sum, we can have:

$15 \div 30 = (15 \cdot 3) \div (30 \cdot 3) = (15 \div 3) \div (30 \div 3)$, that is, $\dfrac{15}{30} = \dfrac{15 \cdot 3}{30 \cdot 3} = \dfrac{\frac{15}{3}}{\frac{30}{3}}$.

Next, moving on to another basic law, we have:

- If we multiply (or divide) by the same both sides of an equality, the equality maintains.

Suppose for instance, A, B, C, and D are numbers, constants, monomials, polynomials, or expressions in other kinds, and $A = B$.

Then, we can get: $AC = BC$, and $\dfrac{A}{D} = \dfrac{B}{D}$, where $D \neq 0$, of course.

Next, moving on to another basic law, we have:

- If we add the same to both sides of an equality, the equality maintains.

And the same is true for subtractions, too. So we can say that:

- If we subtract the same from both sides of an equality, the equality maintains.

Suppose for instance, A, B, C, and D are numbers, constants, monomials, polynomials, or expressions in other kinds, and $A = B$.

Then, we can get: $A + C = B + C$, and $A - C = B - C$.

In sum, we have: $A \pm C = B \pm C$.

What then, is good about the basic laws above?

We often use the laws, or rather, we have to go by them when solving equations.

Solving for instance, two equations: $2x - 5 = 6$, and $0.5x + 3 = 4$, we can get:

$2x - 5 = 3 \Rightarrow 2x - 5 + 5 = 3 + 5 \Rightarrow 2x = 8 \Rightarrow 2x/2 = 8/2 = 4 \Rightarrow x = 4$. And next, we get:

$0.5x + 3 = 4 \Rightarrow 0.5x + 3 - 3 = 4 - 3 \Rightarrow 0.5x = 1 \Rightarrow 0.5x \cdot 2 = 1 \cdot 2 = 2 \Rightarrow x = 2$.

So doing the same to both sides at each step, we eventually get the solution.

Normally though, in case of adding the same to both sides, we do it this way, of course:
$2x - 5 = 3 \Rightarrow 2x = 3 + 5$.

So it looks as if we moved a term to the other side changing the sign.
What's actually happened is however, we've added the same to both sides.

Next, we have another set of laws often used in arithmetic, and usually call them **three basic laws**, which however, do not always apply in subtractions and divisions. The three laws always apply in additions and multiplications only. And the three are as follows:

One is **communicative**, another is **associative**, and the other is **distributive**.

- **Commutative law:** $A + B = B + A$, and $AB = BA$.

So for instance, we can have:

$1 + 2 = 2 + 1 = 3$, $1 + (-2) = (-2) + 1 = -1$, and $(-1) + (-2) = (-2) + (-1) = -3$.

$x + y = y + x$, $x + (-y) = (-y) + x = x - y$, and $(-x) + (-y) = (-y) + (-x) = -x - y = -(x + y)$.

$3 \cdot 4 = 4 \cdot 3 = 12$, $3 \cdot (-4) = (-4) \cdot 3 = -12$, and $(-3) \cdot (-4) = (-4) \cdot (-3) = 12$.

$xy = yx$, $x(-y) = (-y)x = -xy$, and $(-x)(-y) = (-y)(-x) = xy$.

● **Associative law:** $A + B + C = (A + B) + C = A + (B + C)$, and $ABC = (AB)C = A(BC)$.

So when we do additions only, the order does not matter.
And the same is true for doing multiplications only, too.

So for instance:

$1 + 2 + 3 = (1 + 2) + 3 = 1 + (2 + 3) = 6$

$-1 + (-2) + (-3) = \{-1 + (-2)\} + (-3) = -1 + \{-2 + (-3)\} = -6$

$1 + (-2) + (-3) = \{1 + (-2)\} + (-3) = 1 + \{-2 + (-3)\} = -4$

$-1 + 2 + (-3) = (-1 + 2) + (-3) = -1 + \{2 + (-3)\} = -2$

$1 + 2 + (-3) = (1 + 2) + (-3) = 1 + \{2 + (-3)\} = 0$

$1 + (-2) + 3 = \{1 + (-2)\} + 3 = 1 + (-2 + 3) = 2$

$x + y + z = (x + y) + z = x + (y + z)$

$-x + (-y) + (-z) = \{-x + (-y)\} + (-z) = -x + \{-y + (-z)\} = -x - y - z = -(x + y + z)$

$x + (-y) + (-z) = \{x + (-y)\} + (-z) = x + \{-y + (-z)\} = x - y - z$

$-x + y + (-z) = (-x + y) + (-z) = -x + \{y + (-z)\} = -x + y - z = y - x - z$

$x + y + (-z) = (x + y) + (-z) = x + \{y + (-z)\} = x + y - z$

$x + (-y) + z = \{x + (-y)\} + z = x + (-y + z) = x - y + z$

$2 \cdot 3 \cdot 4 = (2 \cdot 3)4 = 2(3 \cdot 4) = 24$

$-2 \cdot (-3) \cdot (-4) = \{-2 \cdot (-3)\}(-4) = -2\{-3 \cdot (-4)\} = -24$

$2 \cdot (-3) \cdot (-4) = \{2 \cdot (-3)\}(-4) = 2\{-3 \cdot (-4)\} = 24$

$-2 \cdot 3 \cdot (-4) = (-2 \cdot 3)(-4) = -2\{3 \cdot (-4)\} = 24$

$-2 \cdot 3 \cdot 4 = (-2 \cdot 3)4 = -2(3 \cdot 4) = -24$

$2 \cdot (-3) \cdot 4 = \{2 \cdot (-3)\}4 = 2(-3 \cdot 4) = -24$

$x \cdot y \cdot z = (x \cdot y)z = x(y \cdot z) = xyz$

$-x \cdot (-y) \cdot (-z) = \{-x \cdot (-y)\}(-z) = -x\{-y \cdot (-z)\} = -xyz$

$x \cdot (-y) \cdot (-z) = \{x \cdot (-y)\}(-z) = x\{-y \cdot (-z)\} = xyz$

$x \cdot y \cdot (-z) = (x \cdot y)(-z) = x\{y \cdot (-z)\} = -xyz$

$x \cdot (-y) \cdot z = \{x \cdot (-y)\}z = x(-y \cdot z) = -xyz$

$-x \cdot y \cdot (-z) = (-x \cdot y)(-z) = -x\{y \cdot (-z)\} = xyz$

$\{(x^2 - yz)x\}(y + 1) = (x^3 - xyz)(y + 1) = x^3y + x^3 - xy^2z - xyz$, and

$(x^2 - yz)\{x(y + 1)\} = (x^2 - yz)(xy + x) = x^3y + x^3 - xy^2z - xyz$.

So we get: $(x^2 - yz)x(y + 1) = \{(x^2 - yz)x\}(y + 1) = (x^2 - yz)\{x(y + 1)\}$.

• Distributive law: $A(B + C) = AB + AC$, and $A(B - C) = AB - AC$.

So for instance:

$2(3 + 4) = 2 \cdot 3 + 2 \cdot 4 = 6 + 8 = 14$

$-2\{-3 + (-4)\} = -2 \cdot (-3) + (-2) \cdot (-4) = 6 + 8 = 14$

$2\{-3 + (-4)\} = 2 \cdot (-3) + 2 \cdot (-4) = -6 + (-8) = -6 - 8 = -14$

$-2\{3 + (-4)\} = -2 \cdot 3 + (-2) \cdot (-4) = -6 + 8 = 2$

$-2\{(-3) + 4\} = -2 \cdot (-3) + (-2) \cdot 4 = 6 + (-8) = 6 - 8 = -2$

$x(y + z) = xy + xz$, and $-x\{-y + (-z)\} = xy + xz$.

$x\{-y + (-z)\} = -xy + (-xz) = -xy - xz = -x(y + z)$

$-x\{y + (-z)\} = -xy + xz = x(z - y)$

$x\{y + (-z)\} = xy + (-xz) = xy - xz = x(y - z)$

$2(3 - 4) = 2 \cdot 3 - 2 \cdot 4 = 6 - 8 = -2$

$-2\{-3 - (-4)\} = -2 \cdot (-3) - (-2) \cdot (-4) = 6 - 8 = -2$

$2\{-3 - (-4)\} = 2 \cdot (-3) - 2 \cdot (-4) = -6 - (-8) = -6 + 8 = 2$

$-2\{3 - (-4)\} = -2 \cdot 3 - (-2) \cdot (-4) = -6 - 8 = -14$

$-2\{(-3) - 4\} = -2 \cdot (-3) - (-2) \cdot 4 = 6 - (-8) = 6 + 8 = 14$

$x(y - z) = xy - xz$, and $-x\{-y - (-z)\} = xy - xz$.

$x\{-y - (-z)\} = -xy - (-xz) = -xy + xz = -x(-y + z) = x(y - z)$

$-x\{y - (-z)\} = -xy - xz = -x(y + z)$

$x\{y - (-z)\} = xy - (-xz) = xy + xz = x(y + z)$

Besides, we have: $A \cdot B \div C = (A \cdot B) \div C = A \cdot (B \div C)$.

That's because, $A \cdot B \div C = A \cdot B \cdot \dfrac{1}{C} = (A \cdot B) \cdot \dfrac{1}{C} = A \cdot (B \cdot \dfrac{1}{C}) = \dfrac{AB}{C}$.

In fact, we have: $A \cdot B \div C = (A \cdot B) \div C = \dfrac{AB}{C}$, and $A \cdot (B \div C) = A \cdot \dfrac{B}{C} = \dfrac{AB}{C}$.

For instance, $3 \cdot 4 \div 2 = 3 \cdot 4 \cdot \frac{1}{2} = (3 \cdot 4) \cdot \frac{1}{2} = 3 \cdot (4 \cdot \frac{1}{2}) = 6$, and

$-3 \cdot (-4) \div (-2) = -3 \cdot (-4) \cdot (-\frac{1}{2}) = \{-3 \cdot (-4)\} \cdot (-\frac{1}{2}) = -3 \cdot \{-4 \cdot (-\frac{1}{2})\} = -6$.

In cases of <u>subtractions and divisions</u> however, the three basic laws do <u>not always</u> apply.

In divisions and subtractions, operands do not commute or associate. And a division doesn't distribute itself. And let's see now, some examples.

- To begin with, we have: $A \div B \neq B \div A$, that is, $\dfrac{A}{B} \neq \dfrac{B}{A}$, and $A - B \neq B - A$.

For instance, we have: $15 \div 3 \neq 3 \div 15$, that is, $\dfrac{15}{3} \neq \dfrac{3}{15}$, and $15 - 3 \neq 3 - 15$.

- Next, we have: $(A \div B) \div C \neq A \div (B \div C)$, and operands do not associate in divisions.

That's because: $(A \div B) \div C = \frac{\frac{A}{B}}{C} = \frac{A}{B} \cdot \frac{1}{C} = \frac{A}{BC}$, but $A \div (B \div C) = \frac{A}{\frac{B}{C}} = A \cdot \frac{1}{\frac{B}{C}} = A \cdot \frac{C}{B} = \frac{AC}{B}$.

So we have: $A \div B \div C = (A \div B) \div C \neq A \div (B \div C)$.

For instance, we have: $36 \div 6 \div 3 = (36 \div 6) \div 3 \neq 36 \div (6 \div 3)$.

- Also, we have: $(A \div B) \cdot C \neq A \div (B \cdot C)$.

That's because: $A \div B \cdot C = (A \div B) \cdot C = \frac{A}{B} \cdot C = \frac{AC}{B}$, but $A \div B \cdot C \neq A \div (B \cdot C) = \frac{A}{BC}$.

So for instance, we have: $(36 \div 6) \cdot 3 \neq 36 \div (6 \cdot 3)$.

- Next, we have: $A - B - C = (A - B) - C$, but $A - B - C \neq A - (B - C) = A - B + C$.

So we have: $A - B - C = (A - B) - C \neq A - (B - C)$, and operands do not associate in subtractions, either.

For instance, we get: $5 - 3 - 1 = (5 - 3) - 1$, but $5 - 3 - 1 \neq 5 - (3 - 1) = 5 - 3 + 1$.

So we get: $5 - 3 - 1 = (5 - 3) - 1 \neq 5 - (3 - 1)$.

- Next, we have: $A \div (B + C) \neq (A \div B) + (A \div C)$, that is, $\frac{A}{B + C} \neq \frac{A}{B} + \frac{A}{C}$.

So for instance, we have: $24 \div (2 + 4) \neq (24 \div 2) + (24 \div 4)$, that is, $\frac{24}{2 + 4} \neq \frac{24}{2} + \frac{24}{4}$.

- Also, we have: $A \div (B - C) \neq (A \div B) - (A \div C)$, that is, $\dfrac{A}{B-C} \neq \dfrac{A}{B} - \dfrac{A}{C}$.

For instance, we have: $24 \div (4 - 2) \neq (24 \div 4) - (24 \div 2)$, that is, $\dfrac{24}{4-2} \neq \dfrac{24}{4} - \dfrac{24}{2}$.

So a division does not distribute itself.

The four operations in arithmetic are all connected, though.   How?

Doing a subtraction, we are adding the negative. $5 - 3 = 5 + (-3)$.

Doing a division, we are multiplying the reciprocal. $6/2 = 6 \cdot (1/2)$.

And adding together many of the same things, we do a multiplication. $4 + 4 + 4 = 3 \cdot 4$.

1.2. **Polynomial Arithmetic 3**

The four operations in arithmetic are all connected. How?

Arithmetic starts with an addition, which is a reverse operation of a subtraction, which is in fact, an addition of the number negative.

And adding together same numbers, we do a multiplication, which is a reverse operation of a division, which is in fact, the multiplication of the reciprocal.

Let's now, have a look at some more examples on operations in polynomial arithmetic.

To begin with, assuming $A = 2x + 3y$, and $B = 5x + 7y$, we can get:

$A + B = 2x + 3y + 5x + 7y = x(2 + 5) + y(3 + 7) = 7x + 10y$.

$2A - B = 2(2x + 5y) - (5x + 7y) = 4x + 10y - 5x - 7y = x(4 - 5) + y(10 - 7) = -x + 3y$.

$3A + 2y = 3(2x + 5y) + 2y = 6x + 15y + 2y = 6x + 17y$.

$\frac{1}{2}(2A - B) = \frac{1}{2}\{2(2x + 3y) - (5x + 7y)\} = \frac{1}{2}(4x + 6y - 5x - 7y) = \frac{1}{2}(-x - y) = -\frac{1}{2}(x + y)$.

Next, assuming $A = 3xy^2 + yz$, $B = 2x + 7xy^2 + z + 9yz$, & $C = 3x + 8y + 9z$, we can get:

$$\tfrac{1}{3}A + 2B - 3C = \tfrac{1}{3}(3xy + yz) + 2(2x + 7xy^2 + z + 9yz) - 3(3x + 8y + 9z)$$

$$= xy + \tfrac{1}{3}yz + 4x + 14xy^2 + 2z + 18yz - 9x - 24y - 27z$$

$$= 14xy^2 + xy + \tfrac{1+3\cdot18}{3}yz - 5x - 24y - 25z = 14xy^2 + xy + \tfrac{55}{3}yz - 5x - 24y - 25z.$$

Assuming next, $A = x + y$, $B = y + z$, and $C = x + z$, we can get:

$$A + B + C = x + y + y + z + x + z = 2x + 2y + 2z = 2(x + y + z)$$

$$A - B + C = x + y - y - z + x + z = 2x$$

$$A + B - C = x + y + y + z - x - z = 2y$$

$$\text{-}A + B + C = \text{-}x - y + y + z + x + z = 2z$$

$$(A - B + C) + (A + B - C) + (\text{-}A + B + C) = 2x + 2y + 2z = 2(x + y + z) = A + B + C$$

$$A + (B + C) = x + y + (y + z + x + z) = x + y + y + z + x + z = 2(x + y + z) = A + B + C$$

$$(A + B) + C = (x + y + y + z) + x + z = x + y + y + z + x + z = 2(x + y + z) = A + B + C$$

$$AB = (x + y)(y + z) = (x + y)B = xB + yB = x(y + z) + y(y + z) = xy + xz + y^2 + yz$$

$$= A(y + z) = Ay + Az = (x + y)y + (x + y)z = xy + y^2 + xz + yz$$

$$(AB)C = (x + y)(y + z)C = (xy + xz + y^2 + yz)C = (xy + xz + y^2 + yz)(x + z)$$

$$= x(xy + xz + y^2 + yz) + z(xy + xz + y^2 + yz)$$

$$= x^2y + x^2z + xy^2 + xyz + xyz + xz^2 + y^2z + yz^2 = (y + z)x^2 + (y^2 + 2yz + z^2)x + y^2z + yz^2.$$

$$A(BC) = A\{(y + z)(x + z)\} = A(xy + yz + xz + z^2) = (x + y)(xy + yz + xz + z^2)$$

$$= x(xy + yz + xz + z^2) + y(xy + yz + xz + z^2)$$

$$= x^2y + xyz + x^2z + xz^2 + xy^2 + y^2z + xyz + yz^2 = x^2y + x^2z + 2xyz + xz^2 + xy^2 + y^2z + yz^2$$

$$= (y + z)x^2 + (y^2 + 2yz + z^2)x + y^2z + yz^2.$$

Assuming next, $A = x - y$, and $B = y - z$, we can get:

$$A + B = (x - y) + (y - z) = x - y + y - z = x - z$$

$$B + A = (y - z) + (x - y) = y - z + x - y = x - z$$

$$AB - BA = (x - y)(y - z) - (y - z)(x - y) = xy - xz - y^2 + yz - (xy - y^2 - xz + yz)$$

$$= xy - xz - y^2 + yz - xy + y^2 + xz - yz = 0.$$

Suppose this time, $A = x + y$, $B = y - z$, and $C = z - x$. Then, we can get:

$$A(B + C) = (x + y)\{(y - z) + (z - x)\} = (x + y)(y - z + z - x) = (x + y)(y - x)$$

$$= xy - x^2 + y^2 - xy = y^2 - x^2, \text{ and}$$

$$AB + AC = (x + y)(y - z) + (x + y)(z - x) = xy - xz + y^2 - yz + xz - x^2 + yz - xy = y^2 - x^2.$$

• And let's next, take a look at how we can do *divisions with polynomials*.

Suppose for instance, $A = 2x^4 + 3x^3 + 5x^2 + 7x$, $B = 8x$, and we want to divide A by B.

Then, we can get: $\frac{A}{B} = \frac{2x^4+3x^3+5x^2+7x}{8x} = \frac{2x^4}{8x} + \frac{3x^3}{8x} + \frac{5x^2}{8x} + \frac{7x}{8x} = \frac{2x^3}{8} + \frac{3x^2}{8} + \frac{5x}{8} + \frac{7}{8}$

$= \frac{1}{4}x^3 + \frac{3}{8}x^2 + \frac{5}{8}x + \frac{7}{8}$, which can be put this way, too: $\frac{1}{8}(2x^3 + 3x^2 + 5x + 7)$.

However, divisions with polynomials are not normally that simple as above.
Basically, doing a division, we get the number of the divisors that the dividend can have.
More precisely, we get the maximum number of the divisors that the dividend can have.
What do we mean by the *divisors* though?

Saying a *divisor* of a particular *number*, we mean an *integer* that divides the particular number. And if a number is said to divide a particular number, the division produces no remainder, that is, the remainder is 0. What then, can we think the particular number is?

The particular number is a multiple of the divisor.
So the particular number is an integer, too, since the divisor is an integer.

Saying however, we divide A by B, we call A the dividend, and call B the divisor.

And in that case, it's not always the case where the division of A by B produces no remainder. That is, the remainder can be nonzero.
So it is not necessarily the case where A is exactly a multiple of B.

So dividing A by B, we get the *maximum number* of Bs that the dividend A can have.
Then, the *maximum number* is called the *quotient*, and the remainder can be 0 or nonzero.

Normally though, we don't say the maximum number of Bs that the dividend A can have, but we just say the number of Bs that the dividend A can have.

And if the remainder exists, that is, the remainder is not 0, A is not a multiple of B.

If the remainder does not exist, that is, the remainder is 0, A is a multiple of B.

So for instance, dividing $(3x^2 + 6y)$ by $(x^2 + 2y)$, we get 3 as the quotient, since the number of the divisors, that is, the number of $(x^2 + 2y)$s that the dividend $(3x^2 + 6y)$ can have is 3.   How come?

> We have: $3x^2 + 6y = (x^2 + 2y) + (x^2 + 2y) + (x^2 + 2y)$.
>
> That is, we have: $3x^2 + 6y = 3 \cdot (x^2 + 2y)$.

And also, dividing $3x^2 + 6y + x + 1$ by $x^2 + 2y$, we get 3 as the quotient, too, since the number of $(x^2 + 2y)$s that the dividend $(3x^2 + 6y + x + 1)$ can have is 3.

And in this case, the remainder is not 0, and is: $x + 1$.   How come?

> We have: $3x^2 + 6y + x + 1 = (x^2 + 2y) + (x^2 + 2y) + (x^2 + 2y) + x + 1$.
>
> That is, we have: $3x^2 + 6y + x + 1 = 3 \cdot (x^2 + 2y) + (x + 1)$. And $x + 1$ is the remainder.

So we do multiplications, too, doing divisions.
Doing a division though, we do not just find the number of the divisors that the dividend can have.
Doing a division in fact, we find first, the product of the quotient and the divisor, and then, the sum of the product and the remainder.

> So for instance, dividing A by B, and assuming Q is the quotient, and R is the remainder, we can put the dividend A the way as follows: $A = BQ + R$.

So for instance, dividing $0.3x^2 + 0.6y + x + 1$ by $x^2 + 2y$, we get 0.3 as the quotient, and get $x + 1$ as the remainder. And thus, we can put the dividend the way below:

$0.3x^2 + 0.6y + x + 1 = 0.3(x^2 + 2y) + x + 1$.

So the dividend can be expressed by the sum of the remainder and the product of the quotient and the divisor.

And of course, the quotient does not have to be a number.
In fact, if a polynomial gets divided by a polynomial, the quotient can be not only a number but a constant, monomial, or polynomial, too.

And doing polynomial divisions, we often do a special division called a *synthetic division*, which is a division of a polynomial by a binomial as $x + 1$, and is quite important. So we are going to cover such a division in a separate section.

We often use the division doing polynomial factorizations. We can use it, too, though, just dividing a polynomial by a binomial. And it is covered in fact, in the next section.

We are going to begin with though, *how a synthetic division works*, and will cover a bit general case where a polynomial gets divided by a binomial not as $x - a$ but as $ax - b$, where a and b are constant, of course.

That's because only if knowing how it generally works, you can actually see how it works, and improve a lot your skill of algebra.
So in the next section, we are going to begin with an example polynomial not specific but a bit general, and the example will be as follows:

$P = (ax - b)Q + R$, where a and b are constant but $a \neq 0$, and $P = c_0x^3 + c_1x^2 + c_2x + c_3$, where c_k is constant for $k = 0, 1, 2$, and 3.

Then, dividing the polynomial P by a binomial $(ax - b)$, we get:

$Q = \frac{1}{a}\{c_0x^2 + (c_1 + c_0\alpha)x + c_2 + c_1\alpha + c_0\alpha^2\}$, which is the quotient, which is a polynomial, too, of course, and $R = c_3 + c_2\alpha + c_1\alpha^2 + c_0\alpha^3$ where $\alpha = \frac{b}{a}$, and R is the remainder.

And of course, if the remainder R is 0, the binomial $(ax - b)$ divides the polynomial P, and of course, so does the quotient Q. It's simply because if $R = 0$, we get: $P = (ax - b)Q$. So $(ax - b)$ and Q both are divisors of P.

And we will get to see the details on the division itself in the next section.

1.3. Polynomial Arithmetic 4

Doing polynomial divisions, we often do a special division called a *synthetic division*, which is a division of a polynomial by a binomial as $x + 1$.

So for instance, dividing $x^3 + 2x^2 - 3x + 2$ by $x - 3$, we can do a synthetic division.

We often use synthetic divisions doing polynomial factorizations.
And we can use it, too, of course, just dividing a polynomial by a binomial. It is in fact, convenient to do such a division by synthetic division.

What we are going to see in this section is however, not a specific case but a case a bit general, and thus, we will cover *how a synthetic division works*.

So in this section, we are going to begin with a polynomial not specific but a bit general, and the polynomial is as follows:

$P = (ax - b)Q + R$, where a and b are constant but $a \neq 0$, and $P = c_0 x^3 + c_1 x^2 + c_2 x + c_3$, where c_k is constant for $k = 0, 1, 2$, and 3.

Then, dividing the polynomial P by the polynomial $(ax - b)$, we get:

$Q = \frac{1}{a}\{c_0 x^2 + (c_1 + c_0 \alpha)x + c_2 + c_1 \alpha + c_0 \alpha^2\}$, which is the quotient, which is a polynomial

and $R = c_3 + c_2 \alpha + c_1 \alpha^2 + c_0 \alpha^3$ where $\alpha = \frac{b}{a}$, and R is the remainder.   How come?

Suppose now, $T = b_0 x^2 + b_1 x + b_2$, where b_k is constant for $k = 0, 1$, and 2.

Suppose also, $P = (x - \alpha)T + r$, where α is constant.

Then, if we divide P by $x - \alpha$, T is the quotient, and r is the remainder, we get:

$$P = c_0x^3 + c_1x^2 + c_2x + c_3 = (x - \alpha)T + r = (x - \alpha)(b_0x^2 + b_1x + b_2) + r$$

$$= b_0x^3 + (b_1 - b_0\alpha)x^2 + (b_2 - b_1\alpha)x + (r - b_2\alpha).$$

And comparing terms in P, we get: $c_0 = b_0$, $c_1 = b_1 - b_0\alpha$, $c_2 = b_2 - b_1\alpha$, and $c_3 = r - b_2\alpha$. So next, we can get:

$b_0 = c_0$.

$b_1 = c_1 + b_0\alpha = c_1 + c_0\alpha \Rightarrow b_1 = c_1 + c_0\alpha$.

$b_2 = c_2 + b_1\alpha = c_2 + c_1\alpha + c_0\alpha^2 \Rightarrow b_2 = c_2 + c_1\alpha + c_0\alpha^2$.

$r = c_3 + b_2\alpha = c_3 + c_2\alpha + c_1\alpha^2 + c_0\alpha^3 \Rightarrow r = c_3 + c_2\alpha + c_1\alpha^2 + c_0\alpha^3$.

And we have set: $P = (x - \alpha)T + r$ where $\alpha = \frac{b}{a}$. So we can put P the way below, too:

$$P = a \cdot \frac{1}{a}(x - \alpha)T + r = \frac{1}{a}(ax - a\alpha)T + r = (ax - a\frac{b}{a}) \cdot \frac{1}{a} \cdot T + r = (ax - b) \cdot \frac{1}{a} \cdot T + r.$$

And also, we have: $T = b_0x^2 + b_1x + b_2$, too.

So we get: $P = (ax - b) \cdot \frac{1}{a} \cdot T + r = (ax - b) \cdot \frac{1}{a} \cdot (b_0x^2 + b_1x + b_2) + r$.   Besides, we have:

$b_0 = c_0$, $b_1 = c_1 + c_0\alpha$, $b_2 = c_2 + c_1\alpha + c_0\alpha^2$, and $r = c_3 + c_2\alpha + c_1\alpha^2 + c_0\alpha^3$.

So doing substitutions for b_k where $k = 0$, 1, and 2. we get:

$$P = (ax - b)\frac{1}{a}\{c_0x^2 + (c_1 + c_0\alpha)x + (c_2 + c_1\alpha + c_2\alpha^2)\} + (c_3 + c_2\alpha + c_1\alpha^2 + c_0\alpha^3).$$

Thus now, getting back to this: $P = (ax - b)Q + R$, we get:

$$Q = \frac{1}{a}\{c_0x^2 + (c_1 + c_0\alpha)x + c_2 + c_1\alpha + c_0\alpha^2\}, \text{ and } R = c_3 + c_2\alpha + c_1\alpha^2 + c_0\alpha^3.$$

What if we want to divide such a polynomial as P by just a simple polynomial as $x - 2$?

We can put P in terms of $(x - \alpha)$, and we can simply put it the way below:

$$P = (ax - b)\tfrac{1}{a}\{c_0 x^2 + (c_1 + c_0\alpha)x + c_2 + c_1\alpha + c_0\alpha^2\} + c_3 + c_2\alpha + c_1\alpha^2 + c_0\alpha^3$$

$$= (x - \tfrac{b}{a})\{c_0 x^2 + (c_1 + c_0\alpha)x + c_2 + c_1\alpha + c_0\alpha^2\} + c_3 + c_2\alpha + c_1\alpha^2 + c_0\alpha^3.$$

And we know: $\alpha = \tfrac{b}{a}$.

So we get: $P = (x - \alpha)\{c_0 x^2 + (c_1 + c_0\alpha)x + c_2 + c_1\alpha + c_0\alpha^2\} + c_3 + c_2\alpha + c_1\alpha^2 + c_0\alpha^3.$

Now, if for instance, we set: $\alpha = 2$, the quotient is: $c_0 x^2 + (c_1 + 2c_0)x + c_2 + 2c_1 + 4c_0$, and we can say that the remainder is: $c_3 + 2c_2 + 4c_1 + 8c_0$.

So setting: $P = (x - d)q + R$, where d is constant, and $P = c_0 x^3 + c_1 x^2 + c_2 x + c_3$, where c_k is constant for $k = 0, 1, 2$, and 3, we can see that:

The quotient is: $q = c_0 x^2 + (c_1 + c_0 d)x + c_2 + c_1 d + c_0 d^2$, and the remainder is:

$R = c_3 + c_2 d + c_1 d^2 + c_0 d^3.$

What if $P = c_0 x^4 + c_1 x^3 + c_2 x^2 + c_3 x + c_4$, though?

Then, we get: $q = c_0 x^3 + (c_1 + c_0 d)x^2 + (c_2 + c_1 d + c_0 d^2)x + c_3 + c_2 d + c_1 d^2 + c_0 d^3$, which is the quotient, and $R = c_4 + c_3 d + c_2 d^2 + c_1 d^3 + c_0 d^4$, which is the remainder.

So the division looks quite complicated, doesn't it?

• We can do the division above using a method called a *synthetic division*, and using the synthetic division, we can divide $P = c_0 x^3 + c_1 x^2 + c_2 x + c_3$ by $(x - d)$ the way below:

d	c_0	c_1	c_2	c_3
		$c_0 d$	$c_1 d + c_0 d^2$	$c_2 d + c_1 d^2 + c_0 d^3$
	c_0	$c_1 + c_0 d$	$c_2 + c_1 d + c_0 d^2$	$c_3 + c_2 d + c_1 d^2 + c_0 d^3$

What's going on?

To begin with, put below the horizontal line, the coefficient c_0 of x^3 in the polynomial P.

Next, multiplying c_0 by d, we get: $c_0 d$, which is put below the coefficient c_1 of x^2 in P.

Then, adding together c_1 and $c_0 d$, we get: $c_1 + c_0 d$, which is put below the horizontal line.

Next, multiplying $(c_1 + c_0 d)$ by d, we get: $c_1 d + c_0 d^2$, which is put below the coefficient c_2 of x in P.

Then, adding together c_2 and $(c_1 d + c_0 d^2)$, we get: $c_2 + c_1 d + c_0 d^2$, which is put below the horizontal line.

Next, multiplying $(c_2 + c_1 d + c_0 d^2)$ by d, we get: $c_2 d + c_1 d^2 + c_0 d^3$, which is put below the constant term c_3 in P.

Then, adding together c_3 and $(c_2 d + c_1 d^2 + c_0 d^3)$, we get: $c_3 + c_2 d + c_1 d^2 + c_0 d^3$, which is put below the horizontal line, and is the remainder.

So setting: $P = (x - d)q + R$, we can say that q is the quotient and that R is the remainder, and reflecting the facts stated in the steps above, we can say that:

The first term c_0 below the horizontal line is the coefficient of x^2 in the quotient q.

The second term $(c_1 + c_0 d)$ below the line is the coefficient of x in q.

The third term $(c_2 + c_1 d + c_0 d^2)$ below the line is the constant term in q.

That is, $q = c_0 x^2 + (c_1 + c_0 d)x + c_2 + c_1 d + c_0 d^2$, and $R = c_3 + c_2 d + c_1 d^2 + c_0 d^3$.

And of course, if the remainder R is 0, the binomial $(x - d)$ divides the polynomial P, and of course, so does the quotient q. It's simply because if $R = 0$, we get: $P = (x - d)q$. So $(x - d)$ and q both are divisors of P.   So what?

Using the fact, we can solve many equations as $5x^3 + 2x^2 - x - 6 = 0$.

We are going to cover the actual steps in the synthetic division, and take a specific example in the next section.

1.4. Polynomial Arithmetic 5

In this section, we are going to take actual steps in the *synthetic division* where we divide a polynomial $P = c_0 x^3 + c_1 x^2 + c_2 x + c_3$ by a binomial $x - d$.

To begin with, dividing P by $x - d$, where d is constant, and assuming q is the quotient, and R is the remainder, we can set: $P = (x - d)q + R$.

Then, we get: $q = c_0 x^2 + (c_1 + c_0 d)x + c_2 + c_1 d + c_0 d^2$, and $R = c_3 + c_2 d + c_1 d^2 + c_0 d^3$.

And of course, if the remainder R is 0, the binomial $(x - d)$ divides the polynomial P, and of course, so does the quotient q. It's simply because if $R = 0$, we get: $P = (x - d)q$. So $(x - d)$ and q both are divisors of P.

And in this case, if $x = d$, we get: $P = 0$, since $(x - d)q = (d - d)q = 0$.

So using the fact above, we can solve many equations as $x^3 - 2x^2 + 2x - 1 = 0$.
In fact, $x - 1$ divides $x^3 - 2x^2 + 2x - 1$, because $1^3 - 2 \cdot 1^2 + 2 - 1 = 1 - 2 + 2 - 1 = 0$.
And we can set: $x^3 - 2x^2 + 2x - 1 = (x - 1)q = 0$, where q is a polynomial of degree 2.

And finding such a polynomial as the quotient q, we can use a method called a *synthetic division*, which is shown below.

d	c_0	c_1	c_2	c_3
		$c_0 d$	$c_1 d + c_0 d^2$	$c_2 d + c_1 d^2 + c_0 d^3$
	c_0	$c_1 + c_0 d$	$c_2 + c_1 d + c_0 d^2$	$c_3 + c_2 d + c_1 d^2 + c_0 d^3$

And doing the division the way above, we can take the steps the way below.

To begin with, put below the horizontal line, the coefficient c_0 of x^3 in the polynomial P.

d	c_0	c_1	c_2	c_3
	c_0			

Next, multiplying c_0 by d, we get: $c_0 d$, which is put below the coefficient c_1 of x^2 in P.

d	c_0	c_1	c_2	c_3
		$c_0 d$		
	c_0			

Then, adding together c_1 and $c_0 d$, we get: $c_1 + c_0 d$, which is put below the horizontal line.

d	c_0	c_1	c_2	c_3
		$c_0 d$		
	c_0	$c_1 + c_0 d$		

Next, multiplying $(c_1 + c_0 d)$ by d, we get: $c_1 d + c_0 d^2$, which is put below the coefficient c_2 of x in P.

d	c_0	c_1	c_2	c_3
		$c_0 d$	$c_1 d + c_0 d^2$	
	c_0	$c_1 + c_0 d$		

Then, adding together c_2 and $(c_1 d + c_0 d^2)$, we get: $c_2 + c_1 d + c_0 d^2$, which is put below the horizontal line.

d	c_0	c_1	c_2	c_3
		$c_0 d$	$c_1 d + c_0 d^2$	
	c_0	$c_1 + c_0 d$	$c_2 + c_1 d + c_0 d^2$	

Next, multiplying $(c_2 + c_1 d + c_0 d^2)$ by d, we get: $c_2 d + c_1 d^2 + c_0 d^3$, which is put below the constant term c_3 in P.

d	c_0	c_1	c_2	c_3
		$c_0 d$	$c_1 d + c_0 d^2$	$c_2 d + c_1 d^2 + c_0 d^3$
	c_0	$c_1 + c_0 d$	$c_2 + c_1 d + c_0 d^2$	

Then, adding together c_3 and $(c_2 d + c_1 d^2 + c_0 d^3)$, we get: $c_3 + c_2 d + c_1 d^2 + c_0 d^3$, which is put below the horizontal line, and is the remainder.

d	c_0	c_1	c_2	c_3
		$c_0 d$	$c_1 d + c_0 d^2$	$c_2 d + c_1 d^2 + c_0 d^3$
	c_0	$c_1 + c_0 d$	$c_2 + c_1 d + c_0 d^2$	$c_3 + c_2 d + c_1 d^2 + c_0 d^3$

So the quotient q is as follows:

The first term c_0 below the horizontal line is the coefficient of x^2 in q.

The second term $(c_1 + c_0 d)$ below the line is the coefficient of x in q.

The third term $(c_2 + c_1 d + c_0 d^2)$ below the line is the constant term in q.

That is, $q = c_0 x^2 + (c_1 + c_0 d)x + c_2 + c_1 d + c_0 d^2$, and $R = c_3 + c_2 d + c_1 d^2 + c_0 d^3$.

So assuming: $P = c_0 x^3 + c_1 x^2 + c_2 x + c_3$, dividing P by $x - d$, where d is constant, and assuming q is the quotient, and R is the remainder, we get: $P = (x - d)q + R$, where q and R are as above.

And putting together the entire steps in a sequence, we can put the sequence the way below:

$$
\begin{array}{c|cccc}
d & c_0 & c_1 & c_2 & c_3 \\
\hline
& c_0
\end{array}
$$

$$
\begin{array}{c|cccc}
d & c_0 & c_1 & c_2 & c_3 \\
& & c_0 d & & \\
\hline
& c_0
\end{array}
$$

$$
\begin{array}{c|cccc}
d & c_0 & c_1 & c_2 & c_3 \\
& & c_0 d & & \\
\hline
& c_0 & c_1 + c_0 d
\end{array}
$$

$$
\begin{array}{c|cccc}
d & c_0 & c_1 & c_2 & c_3 \\
& & c_0 d & c_1 d + c_0 d^2 & \\
\hline
& c_0 & c_1 + c_0 d
\end{array}
$$

$$
\begin{array}{c|cccc}
d & c_0 & c_1 & c_2 & c_3 \\
& & c_0 d & c_1 d + c_0 d^2 & \\
\hline
& c_0 & c_1 + c_0 d & c_2 + c_1 d + c_0 d^2
\end{array}
$$

$$
\begin{array}{c|cccc}
d & c_0 & c_1 & c_2 & c_3 \\
& & c_0 d & c_1 d + c_0 d^2 & c_2 d + c_1 d^2 + c_0 d^3 \\
\hline
& c_0 & c_1 + c_0 d & c_2 + c_1 d + c_0 d^2
\end{array}
$$

$$
\begin{array}{c|cccc}
d & c_0 & c_1 & c_2 & c_3 \\
& & c_0 d & c_1 d + c_0 d^2 & c_2 d + c_1 d^2 + c_0 d^3 \\
\hline
& c_0 & c_1 + c_0 d & c_2 + c_1 d + c_0 d^2 & c_3 + c_2 d + c_1 d^2 + c_0 d^3
\end{array}
$$

Let's now, for instance, divide $P = 2x^4 + 3x^3 + 5x^2 + 7x$ by $x - 2$.

Then, setting: $P = (x - d)q + R$, we have: $d = 2$, $c_0 = 2$, $c_1 = 3$, $c_2 = 5$, $c_3 = 7$, and $c_4 = 0$.

So doing the synthetic division, we get the steps as below:

2	2	3	5	7	0
	2				

2	2	3	5	7	0
		4			
	2	7			

2	2	3	5	7	0
		4	14		
	2	7			

2	2	3	5	7	0
		4	14		
	2	7	19		

2	2	3	5	7	0
		4	14	38	
	2	7	19		

2	2	3	5	7	0
		4	14	38	
	2	7	19	45	

2	2	3	5	7	0
		4	14	38	90
	2	7	19	45	90

So the quotient is: $2x^3 + 7x^2 + 19x + 45$, and the remainder is: **90**.

That is, $2x^4 + 3x^3 + 5x^2 + 7x = (x - 2)(2x^3 + 7x^2 + 19x + 45) + 90$.

Let's next, for another instance, divide $P = 2x^4 + 3x^3 + 5x^2 + 7x$ by $x + 2$.

Then, we have: $d = -2$, $c_0 = 2$, $c_1 = 3$, $c_2 = 5$, $c_3 = 7$, and $c_4 = 0$.

So doing the synthetic division, we get the steps as below:

-2	2	3	5	7	0
	2				

-2	2	3	5	7	0
		-4			
	2	-1			

-2	2	3	5	7	0
		-4	2		
	2	-1			

-2	2	3	5	7	0
		-4	2		
	2	-1	7		

-2	2	3	5	7	0
		-4	2	-14	
	2	-1	7		

-2	2	3	5	7	0
		-4	2	-14	14
	2	-1	7	-7	14

So the quotient is: $2x^3 - x^2 + 7x - 7$, and the remainder is **14**.

That is, $2x^4 + 3x^3 + 5x^2 + 7x = (x + 2)(2x^3 - x^2 + 7x - 7) + 14$.

Suppose next, $A = 7x^8 - 4x^7 + 9x^5 + 12x^3 + 14$, $B = 3x^3 + 2x + 1$, and we want to divide A by B. Then, of course, we can put that division this way: $\frac{A}{B} = \frac{7x^8 - 4x^7 + 9x^5 + 12x^3 + 14}{3x^3 + 2x + 1}$. How do we get the quotient and the remainder, though?

Note:
The degree of a polynomial as A above is the largest exponent to which the variable x is raised. So for instance, $2x^8 - 3x^2 + 1$ is a polynomial of degree 8.

Doing a division in numbers, we normally begin with eliminating the highest digit in the dividend, and move on to the next lower digit.
The same is true for a polynomial, too. So we begin with the removal of the term of the highest degree in the polynomial, which we divide, of course, and then, we move on to the next lower. Thus, among terms in the polynomial, we may want to keep degrees in order from the highest to the lowest so that we do not get confused. Putting 0s in front of the terms of missing degrees if any in the polynomial, we can avoid such a confusion.

Now, the term with the highest degree in the dividend A is $7x^8$, which has to be thus, removed first. On the other hand, the term with the highest degree in the divisor B is $3x^3$.

So in the beginning, we need to multiply the divisor by $\frac{7}{3}x^5$ to remove $7x^8$.

That is, we do this operation first: $(3x^3 + 2x + 1)\frac{7}{3}x^5 = 7x^8 + \frac{14}{3}x^6 + \frac{7}{3}x^5$, which is thus, the product of the divisor and $\frac{7}{3}x^5$. So the first step is as below:

$$\frac{\frac{7}{3}x^5\qquad\qquad\qquad\qquad\qquad\qquad\qquad}{3x^3 + 2x + 1\ \big|\ 7x^8 - 4x^7 + 0x^6 + 9x^5 + 0x^4 + 12x^3 + 0x^2 + 0x + 14}$$
$$7x^8 + \tfrac{14}{3}x^6 + \tfrac{7}{3}x^5$$

Removing the term $7x^8$ from the polynomial A, we subtract $7x^8 + \frac{14}{3}x^6 + \frac{7}{3}x^5$ from A.

That is: $A - (7x^8 + \frac{14}{3}x^6 + \frac{7}{3}x^5) = 7x^8 - 4x^7 + 9x^5 + 12x^3 + 14 - (7x^8 + \frac{14}{3}x^6 + \frac{7}{3}x^5)$

$= -4x^7 - \frac{14}{3}x^6 + (9 - \frac{7}{3})x^5 + 12x^3 + 14 = -4x^7 - \frac{14}{3}x^6 + \frac{20}{3}x^5 + 12x^3 + 14$. So we get:

$$\frac{\frac{7}{3}x^5\qquad\qquad\qquad\qquad\qquad\qquad\qquad}{3x^3 + 2x + 1\ \big|\ 7x^8 - 4x^7 + 0x^6 + 9x^5 + 0x^4 + 12x^3 + 0x^2 + 0x + 14}$$
$$\frac{7x^8 + 0x^7 + \tfrac{14}{3}x^6 + \tfrac{7}{3}x^5\qquad\qquad\qquad\qquad\qquad}{-4x^7 - \tfrac{14}{3}x^6 + \tfrac{20}{3}x^5 + 12x^3 + 14}$$

In the next step, assuming: $D = -4x^7 - \frac{14}{3}x^6 + \frac{20}{3}x^5 + 12x^3 + 14,$ we want to divide D by the divisor $3x^3 + 2x + 1$.

Then, we want to remove $-4x^7$ from D.

In other words, we want to multiply the divisor by $-\frac{4}{3}x^4$ to produce $-4x^7$.

That is, we want to get: $(3x^3 + 2x + 1)(-\frac{4}{3}x^4) = -4x^7 - \frac{8}{3}x^5 - \frac{4}{3}x^4.$

So removing the term $-4x^7$ from D, we subtract $(-4x^7 - \frac{8}{3}x^5 - \frac{4}{3}x^4)$ from D.
Then, we get:

$$D - (-4x^7 - \tfrac{8}{3}x^5 - \tfrac{4}{3}x^4) = -4x^7 - \tfrac{14}{3}x^6 + \tfrac{20}{3}x^5 + 12x^3 + 14 - (-4x^7 - \tfrac{8}{3}x^5 - \tfrac{4}{3}x^4)$$

$$= -\tfrac{14}{3}x^6 + \tfrac{20+8}{3}x^5 + \tfrac{4}{3}x^4 + 12x^3 + 14 = -\tfrac{14}{3}x^6 + \tfrac{28}{3}x^5 + \tfrac{4}{3}x^4 + 12x^3 + 14.$$

So we get:

$$
\begin{array}{r}
\frac{7}{3}x^5 - \frac{4}{3}x^4 \\
3x^3 + 2x + 1 \enclose{longdiv}{7x^8 - 4x^7 + 0x^6 + 9x^5 + 0x^4 + 12x^3 + 0x^2 + 0x + 14} \\
\underline{7x^8 + 0x^7 + \frac{14}{3}x^6 + \frac{7}{3}x^5} \\
-4x^7 - \frac{14}{3}x^6 + \frac{20}{3}x^5 + 12x^3 + 14 \\
\underline{-4x^7 + 0x^6 - \frac{8}{3}x^5 - \frac{4}{3}x^4} \\
-\frac{14}{3}x^6 + \frac{28}{3}x^5 + \frac{4}{3}x^4 + 12x^3 + 14
\end{array}
$$

Then, we repeat the same process.

So assuming next, $E = -\frac{14}{3}x^6 + \frac{28}{3}x^5 + \frac{4}{3}x^4 + 12x^3 + 14,$ we want to divide E by the divisor $3x^3 + 2x + 1$. Then, we want to remove $-\frac{14}{3}x^6$ from E. That is, we want to multiply the divisor by $-\frac{14}{9}x^3$ to produce $-\frac{14}{3}x^6$.

So we want to get this time: $(3x^3 + 2x + 1)(-\frac{14}{9}x^3) = -\frac{14}{3}x^6 - \frac{24}{9}x^4 - \frac{14}{9}x^3.$

Removing the term $-\frac{14}{3}x^6$ from E, we subtract $-\frac{14}{3}x^6 - \frac{24}{9}x^4 - \frac{14}{9}x^3$ from E. So we get:

$$E - (-\tfrac{14}{3}x^6 - \tfrac{24}{9}x^4 - \tfrac{14}{9}x^3) = -\tfrac{14}{3}x^6 + \tfrac{28}{3}x^5 + \tfrac{4}{3}x^4 + 12x^3 + 14 - (-\tfrac{14}{3}x^6 - \tfrac{24}{9}x^4 - \tfrac{14}{9}x^3)$$

$$= \tfrac{28}{3}x^5 + (\tfrac{4}{3} + \tfrac{28}{9})x^4 + (12 + \tfrac{14}{9})x^3 + 14 = \tfrac{28}{3}x^5 + \tfrac{40}{9}x^4 + \tfrac{122}{9}x^3 + 14.$$

Then, we get:

$$3x^3 + 2x + 1 \overline{\left| 7x^8 - 4x^7 + 0x^6 + 9x^5 + 0x^4 + 12x^3 + 0x^2 + 0x + 14 \right.}$$

with quotient $\frac{7}{3}x^5 - \frac{4}{3}x^4 - \frac{14}{9}x^3$

$$7x^8 + 0x^7 + \tfrac{14}{3}x^6 + \tfrac{7}{3}x^5$$

$$-4x^7 - \tfrac{14}{3}x^6 + \tfrac{20}{3}x^5 + 12x^3 + 14$$

$$-4x^7 + 0x^6 - \tfrac{8}{3}x^5 - \tfrac{4}{3}x^4$$

$$-\tfrac{14}{3}x^6 + \tfrac{28}{3}x^5 + \tfrac{4}{3}x^4 + 12x^3 + 14$$

$$-\tfrac{14}{3}x^6 + 0x^5 - \tfrac{24}{9}x^4 - \tfrac{14}{9}x^3$$

$$\tfrac{28}{3}x^5 + \tfrac{40}{9}x^4 + \tfrac{122}{9}x^3 + 14$$

Now, we can continue with the rest of the divisions in the same manner until the degree of the remainder is less than the degree of the divisor. Then, we get:

$$3x^3 + 2x + 1 \overline{\left| 7x^8 - 4x^7 + 0x^6 + 9x^5 + 0x^4 + 12x^3 + 0x^2 + 0x + 14 \right.}$$

with quotient $\frac{7}{3}x^5 - \frac{4}{3}x^4 - \frac{14}{9}x^3 + \frac{28}{9}x^2 + \frac{40}{27}x + \frac{22}{9}$

$$7x^8 + 0x^7 + \tfrac{14}{3}x^6 + \tfrac{7}{3}x^5$$

$$-4x^7 - \tfrac{14}{3}x^6 + \tfrac{20}{3}x^5 + 12x^3 + 14$$

$$-4x^7 + 0x^6 - \tfrac{8}{3}x^5 - \tfrac{4}{3}x^4$$

$$-\tfrac{14}{3}x^6 + \tfrac{28}{3}x^5 + \tfrac{4}{3}x^4 + 12x^3 + 14$$

$$-\tfrac{14}{3}x^6 + 0x^5 - \tfrac{24}{9}x^4 - \tfrac{14}{9}x^3$$

$$\tfrac{28}{3}x^5 + \tfrac{40}{9}x^4 + \tfrac{122}{9}x^3 + 14$$

$$\tfrac{28}{3}x^5 + 0x^4 + \tfrac{56}{9}x^3 + \tfrac{28}{9}x^2$$

$$\tfrac{40}{9}x^4 + \tfrac{22}{3}x^3 - \tfrac{28}{9}x^2 + 14$$

$$\tfrac{40}{9}x^4 + 0x^3 + \tfrac{80}{27}x^2 + \tfrac{40}{27}x$$

$$\tfrac{22}{3}x^3 - \tfrac{164}{27}x^2 - \tfrac{40}{27}x + 14$$

$$\tfrac{22}{3}x^3 + 0x^2 + \tfrac{44}{9}x + \tfrac{22}{9}$$

$$-\tfrac{164}{27}x^2 - \tfrac{172}{9}x + \tfrac{104}{9}$$

Therefore, we can see that: $A = BQ + R$ where:

$$Q = \tfrac{7}{3}x^5 - \tfrac{4}{3}x^4 - \tfrac{14}{9}x^3 + \tfrac{28}{9}x^2 + \tfrac{40}{27}x + \tfrac{22}{9}, \text{ and } R = -\tfrac{164}{27}x^2 - \tfrac{172}{9}x + \tfrac{104}{9}.$$

That is:

$$7x^8 - 4x^7 + 9x^5 + 12x^3 + 14$$

$$= (3x^3 + 2x + 1)(\tfrac{7}{3}x^5 - \tfrac{4}{3}x^4 - \tfrac{14}{9}x^3 + \tfrac{28}{9}x^2 + \tfrac{40}{27}x + \tfrac{22}{9}) - \tfrac{164}{27}x^2 - \tfrac{172}{9}x + \tfrac{104}{9}.$$

• Let's try another simple example.

Dividing $A = 2x^5 - 3x^4 + 16x^3 - 19x^2 + 40$ by $B = 2x^3 - 3x^2 + 5$, we get:

$$
\begin{array}{r}
x^2 + 8 \\
\hline
2x^3 - 3x^2 + 5 \,\big|\, 2x^5 - 3x^4 + 16x^3 - 19x^2 + 40 \\
\underline{2x^5 - 3x^4 + 0x^3 + 5x^2} \\
16x^3 - 24x^2 + 40 \\
\underline{16x^3 - 24x^2 + 40} \\
0
\end{array}
$$

So we can see that $\dfrac{A}{B} = \dfrac{2x^5 - 3x^4 + 16x^3 - 19x^2 + 40}{2x^3 - 3x^2 + 5} = x^2 + 8.$

In this case, the remainder is 0, so B can be said to divide A, and also, can be said to be a divisor of A. And of course, the quotient ($x^2 + 8$) can be a divisor, too. Thus:

If we have: $P = DQ + R$ where $R \neq 0$, D cannot be said to be a divisor, and neither can Q.

If however, we have: $P = DQ$, D and Q both can be said to be divisors.

So the word, *divisor* can be ambiguous, can't it?

Just dividing P by D, we just often say D is the divisor, and P is the dividend even if the division produces a nonzero remainder.

We can use a different word for a divisor though, and it is called a factor. So factors are divisors, which can divide a dividend with 0 remainder. What divisor then, is a factor?

It is the core of this book, and is covered in the section, "**What is a factorization?**"

By the way, we may want to check out a calculation tool, which is quite often used in fractional arithmetic. And the tool is as below:

$$\frac{C}{AB} = \frac{C}{B-A}\left(\frac{1}{A} - \frac{1}{B}\right),$$ where $A \neq B$, and A and B both $\neq 0$, of course. How come?

To begin with, we can have: $\frac{1}{A} - \frac{1}{B} = \frac{B}{AB} - \frac{A}{AB} = \frac{B-A}{AB}$.

So we get: $\frac{C}{B-A}\left(\frac{1}{A} - \frac{1}{B}\right) = \frac{C}{B-A} \cdot \frac{B-A}{AB} = \frac{C}{AB}$, and thus, $\frac{C}{AB} = \frac{C}{B-A}\left(\frac{1}{A} - \frac{1}{B}\right)$.

For instance, $\frac{2}{21} = \frac{2}{3 \cdot 7} = \frac{2}{7-3}\left(\frac{1}{3} - \frac{1}{7}\right) = \frac{2}{4}\left(\frac{1}{3} - \frac{1}{7}\right) = \frac{1}{2}\left(\frac{1}{3} - \frac{1}{7}\right)$.

$\frac{1}{2 \cdot 3} + \frac{1}{3 \cdot 4} + \frac{1}{4 \cdot 5} + \ \cdots \ + \frac{1}{99 \cdot 100} = \frac{1}{2} - \frac{1}{3} + \frac{1}{3} - \frac{1}{4} + \frac{1}{4} - \frac{1}{5} + \ \cdots \ + \frac{1}{98} - \frac{1}{99} + \frac{1}{99} - \frac{1}{100} = \frac{1}{2} - \frac{1}{100}$.

Examples 1 in Polynomial Arithmetic

Note:

As all the other examples do in this book, the ones here, too, cover not only practices on the topics covered, but some more ideas, and calculation tools and technics, too, which can help us do *algebra* quickly as well as properly.

0. Assuming: $4x^3 - 5x^2 + 7x + 3 = a(x-2)^3 + b(x-2)^2 + c(x-2) + d$, find the values of a, b, c, and d.

1. Simplify $N = \dfrac{3x+4}{x+1} - \dfrac{5x+12}{x+2} + \dfrac{2x+7}{x+3}$.

2. Simplify $N = \dfrac{1}{x(x+2)} + \dfrac{1}{(x+2)(x+4)} + \dfrac{1}{(x+4)(x+6)}$.

3. Find the values of A, B and C satisfying each of the identical equations below.

3.0. $\dfrac{3x}{x^3+1} = \dfrac{3x}{(x+1)(x^2-x+1)} = \dfrac{A}{x+1} + \dfrac{Bx+C}{x^2-x+1}$.

3.1. $\dfrac{1}{x^3+x^2} = \dfrac{1}{x^2(x+1)} = \dfrac{A}{x} + \dfrac{B}{x^2} + \dfrac{C}{x+1}$.

Suggestions or Solutions

To the **Problem** in the Example **0**

Assuming: $4x^3 - 5x^2 + 7x + 3 = a(x-2)^3 + b(x-2)^2 + c(x-2) + d$, **find the values of** a, b, c, **and** d.

$$
\begin{array}{r|rrrr}
2 & 4 & -5 & 7 & 3 \\
  &   & 8 & 6 & 26 \\
\hline
2 & 4 & 3 & 13 & 29 \quad \Rightarrow d = 29. \\
  &   & 8 & 22 & \\
\hline
2 & 4 & 11 & 35 \quad \Rightarrow c = 35. \\
  &   & 8 & \\
\hline
  & 4 & 19 \quad \Rightarrow a = 4, \text{ and } b = 19.
\end{array}
$$

If not quite sure of how we can get the values, follow the steps below:

Let's begin with setting: $P = 4x^3 - 5x^2 + 7x + 3$.

Then, we can set: $P = a(x-2)^3 + b(x-2)^2 + c(x-2) + d$, too.

Then, the problem is saying that using the expression in $P = 4x^3 - 5x^2 + 7x + 3$, we can find the values of a, b, c, and d.   How?

By means of the polynomial $P = a(x-2)^3 + b(x-2)^2 + c(x-2) + d$.

So examining the polynomial $P = a(x-2)^3 + b(x-2)^2 + c(x-2) + d$, we can notice that if $x = 2$, we get $P = d$.   How can we get d though?

Using the expression in $P = 4x^3 - 5x^2 + 7x + 3$, we can find it.   How?

Putting **2** into x in the polynomial P right above, we get:

$P = 4x^3 - 5x^2 + 7x + 3 = 3 \cdot 2^3 - 5 \cdot 2^2 + 7 \cdot 2 + 3 = 32 - 20 + 14 + 3 = 29 \Rightarrow P = 29$ if $x = 2$.

And also, if $x = 2$, we get: $P = d$, too, from $P = a(x-2)^3 + b(x-2)^2 + c(x-2) + d$.
So we get: $d = 29$.   What about the other values, though?

First, we can put: $P = a(x-2)^3 + b(x-2)^2 + c(x-2) + d$ the way below, too:

$$P = a(x-2)^3 + b(x-2)^2 + c(x-2) + d = (x-2)\{a(x-2)^3 + b(x-2)^2 + c\} + d.$$

So next, setting: $Q = a(x-2)^2 + b(x-2) + c$, we get: $P = (x-2)Q + d$.

Then, we can say that if we divide $P = 4x^3 - 5x^2 + 7x + 3$ by $x-2$, Q is the quotient, and d is the remainder.   So let's do the division.

We can do the synthetic division the way below:
(If not sure of how we can do it the way below, refer to **Polynomial Arithmetic 5.**)

$$
\begin{array}{r|rrrr}
2 & 4 & -5 & 7 & 3 \\
  &   & 8 & 6 & 26 \\
\hline
  & 4 & 3 & 13 & 29
\end{array}
$$

And we have: $P = (x-2)Q + d$.
So we can readily see that $Q = 4x^2 + 3x + 13$, and the remainder is 29, which is d.

Then, by the same method we did for the value of d, we can find c using the polynomial $Q = 4x^2 + 3x + 13$, together with $Q = a(x-2)^2 + b(x-2) + c$.

That is, putting 2 into the polynomial $Q = 4x^2 + 3x + 13$, we get: $Q = $ _a particular value_, and putting 2 into x in the polynomial $Q = a(x-2)^2 + b(x-2) + c$, we get: $Q = c$, which is _the particular value_ stated above.

And by the same token, we can find the values of a and b, too, in a sequential manner.

That is, for instance, setting: $R = a(x-2) + b$, we get: $Q = (x-2)R + c$.   How come?

That's because: $Q = a(x-2)^2 + b(x-2) + c = (x-2)\{a(x-2)^2 + b\} + c = (x-2)R + c$.

Next, for instance, setting: $S = a$, we get: $R = (x-2)S + b$, because $R = a(x-2) + b$.

And thus, putting threads together, we can say that:

The division of P by $(x-2)$ produces d as the remainder, the division of Q by $(x-2)$ yields c as the remainder, and dividing R by $(x-2)$, we get a as the quotient and b as the remainder. So we can sequentially do the synthetic division the way below:

$$
\begin{array}{c|cccc}
2 & 4 & -5 & 7 & 3 \\
  &   & 8 & 6 & 26 \\
\hline
2 & 4 & 3 & 13 & 29 \\
  &   & 8 & 22 & \\
\hline
2 & 4 & 11 & 35 & \\
  &   & 8 & & \\
\hline
  & 4 & 19 & &
\end{array}
$$

$\Rightarrow Q = 4x^2 + 3x + 13$, and $d = 29$.

$\Rightarrow R = 4x + 11$, and $c = 35$.

$\Rightarrow a = 4$, and $b = 19$.

That is to say that:

$P = 4x^3 - 5x^2 + 7x + 3 = (x-2)Q + 29 = (x-2)(4x^2 + 3x + 13) + 29$.

$Q = 4x^2 + 3x + 13 = (x-2)R + 35 = (x-2)(4x + 11) + 35$.

$R = 4x + 11 = (x-2)a + 19 = (x-2)4 + 19$.

So we can see that $P = 4(x-2)^3 + 19(x-2)^2 + 35(x-2) + 29$.

By the way, setting: $P(x) = 4(x-2)^3 + 19(x-2)^2 + 35(x-2) + 29$, we get:

$P(x+2) = 4(x+2-2)^3 + 19(x+2-2)^2 + 35(x+2-2) + 29 = 4x^3 + 19x^2 + 35x + 29$.

So we get: $P(x+2) = 4x^3 + 19x^2 + 35x + 29$.

Suggestions or Solutions
To the **Problem** in the Example **1**

Simplify $N = \dfrac{3x+4}{x+1} - \dfrac{5x+12}{x+2} + \dfrac{2x+7}{x+3}$.

$$N = \frac{3x+4}{x+1} - \frac{5x+12}{x+2} + \frac{2x+7}{x+3} = 3 + \frac{1}{x+1} - 5 - \frac{2}{x+2} + 2 + \frac{1}{x+3} = \frac{1}{x+1} - \frac{2}{x+2} + \frac{1}{x+3}.$$

$$= \frac{1}{x+1} - \frac{1}{x+2} - \frac{1}{x+2} + \frac{1}{x+3} = \frac{1}{x+1} - \frac{1}{x+2} - \left(\frac{1}{x+2} - \frac{1}{x+3} \right)$$

$$= \frac{1}{(x+1)(x+2)} - \frac{1}{(x+2)(x+3)} = \frac{1}{x+2} \left(\frac{1}{x+1} - \frac{1}{x+3} \right)$$

$$= \frac{1}{x+2} \cdot \frac{2}{(x+1)(x+3)} = \frac{2}{(x+1)(x+2)(x+3)}.$$

If not quite sure of the idea behind the processes above, follow the steps below:

To begin with, we have: $\dfrac{ma+n}{a} = m + \dfrac{n}{a}$.

So we can get: $\dfrac{3x+4}{x+1} = \dfrac{3x+3+1}{x+1} = \dfrac{3(x+1)+1}{x+1} = 3 + \dfrac{1}{x+1}$.

By the same token, we can get: $\dfrac{5x+12}{x+2} = 5 + \dfrac{2}{x+2}$, and $\dfrac{2x+7}{x+3} = 2 + \dfrac{1}{x+3}$.

Therefore, $N = 3 + \dfrac{1}{x+1} - \left(5 + \dfrac{2}{x+2} \right) + 2 + \dfrac{1}{x+3} = 3 + \dfrac{1}{x+1} - 5 - \dfrac{2}{x+2} + 2 + \dfrac{1}{x+3}$

$$= \frac{1}{x+1} - \frac{2}{x+2} + \frac{1}{x+3}.$$

Also, we have: $\dfrac{1}{x+1}-\dfrac{2}{x+2}+\dfrac{1}{x+3}=\dfrac{1}{x+1}-\dfrac{1}{x+2}-\dfrac{1}{x+2}+\dfrac{1}{x+3}$, and $\dfrac{1}{a}-\dfrac{1}{b}=\dfrac{b-a}{ab}$.

So we get: $N=\dfrac{1}{x+1}-\dfrac{1}{x+2}-\left(\dfrac{1}{x+2}-\dfrac{1}{x+3}\right)=\dfrac{1}{(x+1)(x+2)}-\dfrac{1}{(x+2)(x+3)}$.

And we have this, too: $\dfrac{1}{ac}-\dfrac{1}{bc}=\left(\dfrac{1}{a}-\dfrac{1}{b}\right)\dfrac{1}{c}=\dfrac{1}{c}\left(\dfrac{1}{a}-\dfrac{1}{b}\right)$.

Thus, we get: $N=\dfrac{1}{x+2}\left(\dfrac{1}{x+1}-\dfrac{1}{x+3}\right)=\dfrac{1}{x+2}\cdot\dfrac{2}{(x+1)(x+3)}=\dfrac{2}{(x+1)(x+2)(x+3)}$.

So we can say that: $\dfrac{3x+4}{x+1}-\dfrac{5x+12}{x+2}+\dfrac{2x+7}{x+3}=\dfrac{2}{(x+1)(x+2)(x+3)}$.

In short:

$$N=\frac{3x+4}{x+1}-\frac{5x+12}{x+2}+\frac{2x+7}{x+3}=3+\frac{1}{x+1}-5-\frac{2}{x+2}+2+\frac{1}{x+3}=\frac{1}{x+1}-\frac{2}{x+2}+\frac{1}{x+3}.$$

$$=\frac{1}{x+1}-\frac{1}{x+2}-\frac{1}{x+2}+\frac{1}{x+3}=\frac{1}{x+1}-\frac{1}{x+2}-\left(\frac{1}{x+2}-\frac{1}{x+3}\right)$$

$$=\frac{1}{(x+1)(x+2)}-\frac{1}{(x+2)(x+3)}=\frac{1}{x+2}\left(\frac{1}{x+1}-\frac{1}{x+3}\right)$$

$$=\frac{1}{x+2}\cdot\frac{2}{(x+1)(x+3)}=\frac{2}{(x+1)(x+2)(x+3)}.$$

Suggestions or Solutions
To the **Problem** in the Example **2**

Simplify $N = \dfrac{1}{x(x+2)} + \dfrac{1}{(x+2)(x+4)} + \dfrac{1}{(x+4)(x+6)}.$

To begin with, we have: $\dfrac{1}{a} - \dfrac{1}{b} = \dfrac{b-a}{ab} \Rightarrow \dfrac{1}{ab} = \dfrac{1}{b-a}\left(\dfrac{1}{a} - \dfrac{1}{b}\right).$

So we get: $\dfrac{1}{x(x+2)} = \dfrac{1}{(x+2)-x}\left(\dfrac{1}{x} - \dfrac{1}{x+2}\right) = \dfrac{1}{2}\left(\dfrac{1}{x} - \dfrac{1}{x+2}\right).$

By the same token, we get:

$$\dfrac{1}{(x+2)(x+4)} = \dfrac{1}{2}\left(\dfrac{1}{x+2} - \dfrac{1}{x+4}\right), \text{ and } \dfrac{1}{(x+4)(x+6)} = \dfrac{1}{2}\left(\dfrac{1}{x+4} - \dfrac{1}{x+6}\right).$$

So we get:

$$N = \dfrac{1}{2}\left(\dfrac{1}{x} - \dfrac{1}{x+2}\right) + \dfrac{1}{2}\left(\dfrac{1}{x+2} - \dfrac{1}{x+4}\right) + \dfrac{1}{2}\left(\dfrac{1}{x+4} - \dfrac{1}{x+6}\right) = \dfrac{1}{2}\left(\dfrac{1}{x} - \dfrac{1}{x+6}\right).$$

Then again, since we have: $\dfrac{1}{a} - \dfrac{1}{b} = \dfrac{b-a}{ab}$, we get:

$$N = \dfrac{1}{2}\left(\dfrac{1}{x} - \dfrac{1}{x+6}\right) = \dfrac{1}{2} \cdot \dfrac{(x+6)-x}{x(x+6)} = \dfrac{1}{2} \cdot \dfrac{6}{x(x+6)} = \dfrac{3}{x(x+6)}.$$

In short:

$$N = \dfrac{1}{x(x+2)} + \dfrac{1}{(x+2)(x+4)} + \dfrac{1}{(x+4)(x+6)}$$

$$= \dfrac{1}{2}\left(\dfrac{1}{x} - \dfrac{1}{x+2}\right) + \dfrac{1}{2}\left(\dfrac{1}{x+2} - \dfrac{1}{x+4}\right) + \dfrac{1}{2}\left(\dfrac{1}{x+4} - \dfrac{1}{x+6}\right) = \dfrac{1}{2}\left(\dfrac{1}{x} - \dfrac{1}{x+6}\right) = \dfrac{3}{x(x+6)}.$$

Suggestions or Solutions
To the **Problem 0** in the Example **3**

Find the values of A, B and C satisfying the identical equation below.

$$\frac{3x}{x^3+1} = \frac{3x}{(x+1)(x^2-x+1)} = \frac{A}{x+1} + \frac{Bx+C}{x^2-x+1}.$$

To begin with, $\dfrac{3x}{x^3+1} = \dfrac{3x}{x^3+1} = \dfrac{A(x^2-x+1)+(Bx+C)(x+1)}{x^3+1}$.

So: $3x = A(x^2-x+1) + (x+1)(Bx+C) = x^2(A+B) + x(C+B-A) + A + C$.

Now, $A + B = 0$, and $A + C = 0$, so: $A = -C = -B$.

Thus: $C + B - A = B + B - (-B) = 3B = 3 \Rightarrow B = 1 \Rightarrow A = -1 \Rightarrow C = 1$.

If not quite sure of the idea behind the processes above, follow the steps below:

Examining first, the denominators, we can see those in the first two fractions are the same. How can we so readily and quickly notice that, though?

If much used to polynomial factorizations, we can do so.

Expanding the second one, we get: $(x+1)(x^2-x+1) = x^3 - x^2 + x + x^2 - x + 1 = x^3 + 1$.

Next, taking the common denominator in the fractions on the far right hand side, we get:

$(x+1)(x^2-x+1)$, which is x^3+1. So we get:

$$\frac{A}{x+1} + \frac{Bx+C}{x^2-x+1} = \frac{A(x^2-x+1)}{(x+1)(x^2-x+1)} + \frac{(Bx+C)(x+1)}{(x+1)(x^2-x+1)} = \frac{A(x^2-x+1)+(Bx+C)(x+1)}{(x+1)(x^2-x+1)}.$$

Thus, we get: $\dfrac{3x}{x^3+1} = \dfrac{3x}{x^3+1} = \dfrac{A(x^2-x+1)+(Bx+C)(x+1)}{x^3+1}$.

So we get: $3x = A(x^2-x+1)+(x+1)(Bx+C) = Ax^2 - Ax + A + Bx^2 + Cx + Bx + C$

$= x^2(A+B) + x(-A+C+B) + A + C \Rightarrow 3x = x^2(A+B) + x(C+B-A) + A + C$

$\Rightarrow 0 \cdot x^2 + 3x + 0 = x^2(A+B) + x(C+B-A) + A + C.$

So comparing now, the coefficients and the constant terms in both sides, we get a system of equations as follows: $A + B = 0$, $C + B - A = 3$, and $A + C = 0$.

Now, solving the system above, we can find A, B, and C, of course.

First, $A + B = 0 \Rightarrow A = -B$, and next, $A + C = 0 \Rightarrow A = -C \Rightarrow C = B$ since $A = -B$.

Thus, we get: $C + B - A = B + B - (-B) = 3B = 3 \Rightarrow B = 1 \Rightarrow A = -1 \Rightarrow C = 1$.

So we can put the fraction this way, too: $\dfrac{3x}{x^3+1} = \dfrac{-1}{x+1} + \dfrac{x+1}{x^2-x+1}$.

In short:

To begin with, $\dfrac{3x}{x^3+1} = \dfrac{3x}{x^3+1} = \dfrac{A(x^2-x+1)+(Bx+C)(x+1)}{x^3+1}$.

So: $3x = A(x^2-x+1)+(x+1)(Bx+C) = x^2(A+B) + x(C+B-A) + A + C.$

Now, $A + B = 0$, and $A + C = 0$, so: $A = -C = -B$.

Thus: $C + B - A = B + B - (-B) = 3B = 3 \Rightarrow B = 1 \Rightarrow A = -1 \Rightarrow C = 1.$

Suggestions or Solutions

Find the values of A, B and C satisfying the identical equation below.

$$\frac{1}{x^3+x^2}=\frac{1}{x^2(x+1)}=\frac{A}{x}+\frac{B}{x^2}+\frac{C}{x+1}.$$

To begin with, $\dfrac{1}{x^3+x^2}=\dfrac{1}{x^2(x+1)}=\dfrac{Ax(x+1)+B(x+1)+Cx^2}{x^2(x+1)}.$

So: $1 = Ax(x+1)+B(x+1)+Cx^2 = (A+C)x^2+(A+B)x+B.$

Thus, $A+C=0$, $A+B=0$, and $B=1$.

So: $B=1 \Rightarrow A+B=A+1=0 \Rightarrow A=-1$, and thus, $A+C=-1+C=0 \Rightarrow C=1$.

If not quite sure of the idea behind the processes above, follow the steps below:

Examining the denominators, we can see those in the first two fractions are the same.

Next, taking the common denominator in the fractions on the far right hand side, that is, taking the LCM (the least common multiple) of x, x^2, and $x+1$, we get:

$x^2(x+1)$, which is x^3+x^2. So we get:

$$\frac{A}{x}+\frac{B}{x^2}+\frac{C}{x+1}=\frac{Ax(x+1)}{x^2(x+1)}+\frac{B(x+1)}{x^2(x+1)}+\frac{Cx^2}{x^2(x+1)}=\frac{Ax(x+1)+B(x+1)+Cx^2}{x^2(x+1)}.$$

Thus, we get:

$$1 = Ax(x+1)+B(x+1)+Cx^2 = Ax^2+Ax+Bx+B+Cx^2 = (A+C)x^2+(A+B)x+B.$$

That is, we get: $0{\cdot}x^2+0{\cdot}x+1 = (A+C)x^2+(A+B)x+B.$

So comparing the coefficients and the constant terms in both sides of the equality above, we get a system of equations as follows: $A + C = 0$, $A + B = 0$, and $B = 1$.

Thus, we get: $B = 1 \Rightarrow A + B = A + 1 = 0 \Rightarrow A = \text{-}1$, so $A + C = \text{-}1 + C = 0 \Rightarrow C = 1$.

So we can put the fraction this way, too: $\dfrac{1}{x^3 + x^2} = \dfrac{1}{x^2(x+1)} = \dfrac{\text{-}1}{x} + \dfrac{1}{x^2} + \dfrac{1}{x+1}$.

And we can notice that we can break apart a fraction the way above.

In short:

To begin with, $\dfrac{1}{x^3 + x^2} = \dfrac{1}{x^2(x+1)} = \dfrac{Ax(x+1) + B(x+1) + Cx^2}{x^2(x+1)}$.

So: $1 = Ax(x + 1) + B(x + 1) + Cx^2 = (A + C)x^2 + (A + B)x + B$.

Thus, $A + C = 0$, $A + B = 0$, and $B = 1$.

So we get: $B = 1 \Rightarrow A + B = A + 1 = 0 \Rightarrow A = \text{-}1$, and thus, $A + C = \text{-}1 + C = 0 \Rightarrow C = 1$.

Examples 2 in Polynomial Arithmetic

0.   Find the values of P and Q as follows:

0.0. $P = \dfrac{a}{ab+a+1} + \dfrac{b}{bc+b+1} + \dfrac{c}{ca+c+1}$, and $abc = 1$.

0.1. $Q = \dfrac{a+b}{(a+1)(b+1)} + \dfrac{b+c}{(b+1)(c+1)} + \dfrac{c+a}{(c+1)(a+1)}$, and $abc = -1$.

1.   Assuming that $x + \dfrac{1}{x} = 1$, find the values of P, Q, R, and S as follows:

1.0. $P = x^2 + \dfrac{1}{x^2}$

1.1. $Q = x^3 + \dfrac{1}{x^3}$

1.2. $R = \dfrac{x^{10}+1}{x^2}$

1.3. $S = x^{101} + \dfrac{1}{x^{101}}$

Suggestions or Solutions
To the Problems in the Example 0

Find the values of P and Q as follows:

0.0. $P = \dfrac{a}{ab+a+1} + \dfrac{b}{bc+b+1} + \dfrac{c}{ca+c+1}$, and $abc = 1$.

0.1. $Q = \dfrac{a+b}{(a+1)(b+1)} + \dfrac{b+c}{(b+1)(c+1)} + \dfrac{c+a}{(c+1)(a+1)}$, and $abc = -1$.

0.0. Quite often, reducing the number of variables, we can make expressions simpler.

Using: $abc = 1$, we can do so. So to begin with, we get: $abc = 1 \Rightarrow c = \dfrac{1}{ab}$, and thus:

$$\frac{b}{bc+b+1} = \frac{b}{b \cdot \dfrac{1}{ab} + b + 1} = \frac{b}{\dfrac{1}{a} + b + 1} = \frac{b}{\dfrac{1+ab+a}{a}} = b \cdot \frac{a}{1+ab+a} = \frac{ab}{ab+a+1}.$$

$$\frac{c}{ca+c+1} = \frac{\dfrac{1}{ab}}{a \cdot \dfrac{1}{ab} + \dfrac{1}{ab} + 1} = \frac{\dfrac{1}{ab}}{\dfrac{1}{b} + \dfrac{1}{ab} + 1} = \frac{\dfrac{1}{ab}}{\dfrac{b+1+ab}{ab}} = \frac{1}{ab} \cdot \frac{ab}{b+1+ab} = \frac{1}{ab+b+1}.$$

Thus, we get: $P = \dfrac{a}{ab+a+1} + \dfrac{ab}{ab+a+1} + \dfrac{1}{ab+a+1} = \dfrac{ab+b+1}{ab+a+1} = 1$.

So what is this about?

It's just for algebra practice, because algebra matters. And it does very much so.
Doing algebra, we manipulate math expressions.
Doing algebra very well, we can change or alter expressions keeping their values intact, so we can put the expressions the way we want. What way though?

The way the problems can be solved, of course.

And next, moving on to the problem **0.1**, we get:

$$Q = \frac{a+b}{(a+1)(b+1)} + \frac{b+c}{(b+1)(c+1)} + \frac{c+a}{(c+1)(a+1)}$$

$$= \frac{(a+b)(c+1)}{(a+1)(b+1)(c+1)} + \frac{(a+1)(b+c)}{(a+1)(b+1)(c+1)} + \frac{(b+1)(c+a)}{(b+1)(c+1)(a+1)}$$

$$= \frac{ac+a+bc+b+ab+ac+b+c+bc+ab+c+a}{(a+1)(b+1)(c+1)} = \frac{2ab+2bc+2ca+2a+2b+2c}{(ab+a+b+1)(c+1)}$$

$$= \frac{2(ab+bc+ca+a+b+c)}{(abc+ab+ac+a+bc+b+c+1)} = \frac{2(ab+bc+ca+a+b+c)}{(ab+bc+ca+a+b+c)}, \text{ because } abc = \text{-}1.$$

So we get: $Q = 2$.

Suggestions or Solutions

To the **Problems** in the Example 1

Assuming that $x + \dfrac{1}{x} = 1$, **find the values of** $P, Q, R,$ **and** S **as follows:**

1.0. $P = x^2 + \dfrac{1}{x^2}$ 　　　　　　　**1.1.** $Q = x^3 + \dfrac{1}{x^3}$

1.2. $R = \dfrac{x^{10} + 1}{x^2}$ 　　　　　　**1.3.** $S = x^{101} + \dfrac{1}{x^{101}}$

Since we are given this: $x + \dfrac{1}{x} = 1$, we should be able to take advantage of it.

1.0. $x + \dfrac{1}{x} = 1 \Rightarrow \left(x + \dfrac{1}{x}\right)^2 = \left(x + \dfrac{1}{x}\right)\left(x + \dfrac{1}{x}\right) = x^2 + 1 + 1 + \dfrac{1}{x^2} = 1^2 \Rightarrow x^2 + \dfrac{1}{x^2} = -1.$

1.1. Suppose $a = x$, and $b = \dfrac{1}{x}$.

Then, since $x + \dfrac{1}{x} = 1$, we get: $a + b = 1 \Rightarrow (a + b)^3 = 1$.

Meanwhile:

$(a + b)^3 = (a + b)(a + b)(a + b) = (a^2 + ab + ab + b^2)(a + b) = (a^2 + 2ab + b^2)(a + b)$

$= a^3 + a^2b + 2a^2b + 2ab^2 + ab^2 + b^3 = a^3 + 3a^2b + 3ab^2 + b^3 = a^3 + 3ab(a + b) + b^3 = 1.$

So we now have: $a^3 + 3ab(a + b) + b^3 = 1$.

And we have: $Q = x^3 + \dfrac{1}{x^3} = a^3 + b^3$, $ab = x \cdot \dfrac{1}{x} = 1$, and $a + b = x + \dfrac{1}{x} = 1$.

So we get: $Q + 3 \cdot 1 \cdot 1 = 1 \Rightarrow Q = -2.$

1.2. $R = \dfrac{x^{10}+1}{x^2}$

First, in the problem **1.1**, we found this: $Q = x^3 + \dfrac{1}{x^3} = -2$.

So setting next: $a = x^3$, we can get:

$$a + \frac{1}{a} = -2 \Rightarrow a\left(a + \frac{1}{a}\right) = -2a \Rightarrow a^2 + 1 = -2a \Rightarrow a^2 + 2a + 1 = 0$$

$$\Rightarrow a^2 + a + a + 1 = a(a+1) + (a+1)\cdot 1 = (a+1)(a+1) = (a+1)^2 = 0.$$

So we get: $a + 1 = 0 \Rightarrow a = \text{-}1$, and thus, $x^3 = \text{-}1$, since we've set: $a = x^3$.

Now, $x^{10} = (x^3)^3 x = \text{-}x$ since $x^3 = \text{-}1$, so we get: $R = \dfrac{x^{10}+1}{x^2} = \dfrac{-x+1}{x^2}$. What's next, then?

We have: $x + \dfrac{1}{x} = 1$. So multiplying by x both sides, we can get:

$x^2 + 1 = x \Rightarrow x^2 - x + 1 = 0$. So what does it have to do with this: $R = \dfrac{-x+1}{x^2}$?

We can get: $\text{-}x + 1 = \text{-}x^2$. Thus, we get: $R = \dfrac{-x+1}{x^2} = \dfrac{-x^2}{x^2} = -1.$

1.3. $S = x^{101} + \dfrac{1}{x^{101}}$

In **1.2** above, we found: $x^3 = \text{-}1$. And $101 = 33\cdot 3 + 2$. So we can get: $x^{101} = (x^3)^{33}x^2 = \text{-}x^2$.

Thus, we get: $S = x^{101} + \dfrac{1}{x^{101}} = -x^2 - \dfrac{1}{x^2} = -(x^2 + \dfrac{1}{x^2}) = -(-1)$, because $x^2 + \dfrac{1}{x^2} = -1$,

which is P in the problem **1.0.** above.

60

In short:

1.0. $P = x^2 + \dfrac{1}{x^2}$

$x + \dfrac{1}{x} = 1 \Rightarrow \left(x + \dfrac{1}{x}\right)^2 = x^2 + 2 + \dfrac{1}{x^2} = 1 \Rightarrow P = x^2 + \dfrac{1}{x^2} = -1.$

1.1. $Q = x^3 + \dfrac{1}{x^3}$

Setting $a = x$, and $b = \dfrac{1}{x}$, we get: $a + b = 1$, and $ab = 1$.

So we get: $(a + b)^3 = a^3 + 3ab(a + b) + b^3 = 1 \Rightarrow Q + 3 \cdot 1 \cdot 1 = 1 \Rightarrow Q = -2.$

1.2. $R = \dfrac{x^{10} + 1}{x^2}$

$x + \dfrac{1}{x} = 1 \Rightarrow x^2 + 1 = x \Rightarrow x^2 - x + 1 = 0 \Rightarrow (x + 1)(x^2 - x + 1) = x^3 + 1 = 0 \Rightarrow x^3 = -1.$

So we get: $x^{10} = (x^3)^3 x = -x$, since $x^3 = -1$.

Thus, $R = \dfrac{x^{10} + 1}{x^2} = \dfrac{-x + 1}{x^2} = \dfrac{-x^2}{x^2} = -1$, because $x^2 - x + 1 = 0 \Rightarrow -x + 1 = -x^2.$

1.3. $S = x^{101} + \dfrac{1}{x^{101}}$

We have: $x^3 = -1$, so we get: $x^{101} = (x^3)^{33} x^2 = -x^2.$

Thus: $S = x^{101} + \dfrac{1}{x^{101}} = -x^2 - \dfrac{1}{x^2} = -(x^2 + \dfrac{1}{x^2}) = -(-1)$, since $x^2 + \dfrac{1}{x^2} = -1$, which is

from the problem **1.0.**

Examples 3 in Polynomial Arithmetic

0. Assuming that $x + \dfrac{1}{x} = 4$, find the values of P and Q as follows:

0.0. $P = x - \dfrac{1}{x}$

0.1. $Q = x^3 - \dfrac{1}{x^3}$

1. Assuming that $x^3 - \dfrac{1}{x^3} = 8\sqrt{5}$ where $x > 0$, find the values of P and Q as follows:

1.0. $P = x^3 + \dfrac{1}{x^3}$

1.1. $Q = x + \dfrac{1}{x}$

Suggestions or Solutions
To the Problem in the Example 0

Assuming that $x + \dfrac{1}{x} = 4$, find the values of P and Q as follows:

0.0. $P = x - \dfrac{1}{x}$

0.1. $Q = x^3 - \dfrac{1}{x^3}$

To begin with, letting: $a = x$, and $b = \dfrac{1}{x}$, we get: $a + b = 4$, and $ab = x \cdot \dfrac{1}{x} = 1$.

So next, we can get: $a + b = 4 \Rightarrow (a + b)^2 = a^2 + 2ab + b^2 = a^2 + 2 \cdot 1 + b^2 = 4^2 = 16$

$\Rightarrow a^2 + b^2 = 16 - 2 = 14 \Rightarrow a^2 + b^2 = 14$.

0.0. $P = a - b \Rightarrow P^2 = (a - b)^2 = a^2 - 2ab + b^2 = a^2 + b^2 - 2 \cdot 1 = 14 - 2 = 12 \Rightarrow P^2 = 12$.

Thus, we get: $P = \pm\sqrt{12} = \pm 2\sqrt{3}$.

0.1. Let's begin with setting again, $a = x$, and $b = \dfrac{1}{x}$.

Then first, in the problem **0.0**, we have: $P = \pm 2\sqrt{3}$, and $P = a - b$.

So we can set: $a - b = \pm 2\sqrt{3}$. And thus, we can get:

$(a - b)^3 = (a - b)(a - b)(a - b) = (a^2 - ab - ab + b^2)(a - b) = (a^2 - 2ab + b^2)(a - b)$

$= a^3 - a^2b - 2a^2b + 2ab^2 + ab^2 - b^3 = a^3 - 3a^2b + 3ab^2 - b^3$

$= a^3 - 3ab(a - b) - b^3 = (\pm 2\sqrt{3})^3 = \pm 2^3 3\sqrt{3} = \pm 8 \cdot 3\sqrt{3} = \pm 24\sqrt{3}$.

And we can put it this way, too: $(\pm 2\sqrt{3})^3 = (\pm\sqrt{12})^3 = \pm 12\sqrt{12} = \pm 12\cdot 2\sqrt{3} = \pm 24\sqrt{3}$.

So we now have: $a^3 - 3ab(a - b) - b^3 = \pm 24\sqrt{3}$.

And also, we can have: $ab = 1$, too, because $a = x$, and $b = \dfrac{1}{x}$. So?

So we get: $a^3 - 3ab(a - b) - b^3 = a^3 - 3\cdot 1\cdot(\pm 2\sqrt{3}) - b^3 = a^3 - (\pm 6\sqrt{3}) - b^3 = \pm 24\sqrt{3}$.

And we know: $Q = a^3 - b^3$.

Thus, we get:

$$a^3 - (\pm 6\sqrt{3}) - b^3 = a^3 - b^3 - (\pm 6\sqrt{3}) = Q - (\pm 6\sqrt{3}) = (\pm 24\sqrt{3})$$

$$\Rightarrow Q = (\pm 6\sqrt{3}) + (\pm 24\sqrt{3}) = \pm 30\sqrt{3}.$$

And if knowing polynomial factorizations, we can do it the way below, too:

$$Q = a^3 - b^3 = (a - b)(a^2 + ab + b^2) = P(a^2 + 2ab + b^2 - ab), \text{ since } P = a - b.$$

So we get: $Q = P\{(a + b)^2 - 1\} = P(16 - 1) = 15P$, because we have: $a + b = 4$, since we

have: $x + \dfrac{1}{x} = 4$, and we have set: $a = x$, and $b = \dfrac{1}{x}$.

Thus, we get: $Q = 15P = 15(\pm 2\sqrt{3}) = \pm 30\sqrt{3}$.

And also, we can do it this way, too:

$$P^3 = (a - b)^3 = a^3 - b^3 - 3ab(a - b) = Q - 3\cdot 1\cdot P = Q - 3P.$$

So we get: $Q = P^3 + 3P = P(P^2 + 3) = \pm\sqrt{12}\{(\pm\sqrt{12})^2 + 3\} = \pm 2\sqrt{3}(12 + 3) = \pm 30\sqrt{3}$.

In short:

Assuning first, $a = x$, and $b = \dfrac{1}{x}$, we get: $a + b = 4$, and $ab = x \cdot \dfrac{1}{x} = 1$.

So we get: $(a + b)^2 = a^2 + 2ab + b^2 = a^2 + 2 + b^2 = 4^2 = 16 \Rightarrow a^2 + b^2 = 14$.

0.0. $P = x - \dfrac{1}{x}$

$P^2 = (a - b)^2 = a^2 - 2ab + b^2 = a^2 + b^2 - 2 = 14 - 2 = 12 \Rightarrow P = \pm\sqrt{12} = \pm 2\sqrt{3}$.

0.1. $Q = x^3 - \dfrac{1}{x^3}$

We have: $a - b = \pm 2\sqrt{3}$, which is the value of P above, and $ab = 1$.

So we can get: $(a - b)^3 = a^3 - 3ab(a - b) - b^3 = (\pm 2\sqrt{3})^3 = \pm 24\sqrt{3}$.

Thus, we get: $Q - 3ab(a - b) = Q - 3 \cdot 1 \cdot (\pm 2\sqrt{3}) = Q - (\pm 6\sqrt{3}) = \pm 24\sqrt{3}$.

Therefore, we get: $Q = (\pm 6\sqrt{3}) + (\pm 24\sqrt{3}) = \pm 30\sqrt{3}$.

Suggestions or Solutions
To the **Problem 0** in the Example 1

Assuming that $x^3 - \dfrac{1}{x^3} = 8\sqrt{5}$ where $x > 0$, **find the values of** $P = x^3 + \dfrac{1}{x^3}$.

Setting first: $a = x$, and $b = \dfrac{1}{x}$, we can get: $ab = 1$, and $a^3 - b^3 = 8\sqrt{5}$.

So next, we can get:

$$P = a^3 + b^3 \Rightarrow P^2 = (a^3 + b^3)^2 = (a^3)^2 + 2a^3b^3 + (b^3)^2 = (a^3)^2 - 2a^3b^3 + (b^3)^2 + 4a^3b^3$$

$$= (a^3 - b^3)^2 + 4a^3b^3 = (8\sqrt{5})^2 + 4 = 8^2 \cdot 5 + 4 = 4(2 \cdot 8 \cdot 5 + 1) = 4 \cdot 81 = 18^2.$$

Thus, we get: $P = 18$, because $P > 0$, since $x > 0$.

If not quite sure of the idea behind the processes above, follow the steps below:

Setting first again: $a = x$, and $b = \dfrac{1}{x}$, we can get: $ab = 1$, and $a^3 - b^3 = 8\sqrt{5}$.

So next, we can get:

$$P = a^3 + b^3 \Rightarrow P^2 = (a^3 + b^3)^2 = (a^3 + b^3)(a^3 + b^3) = (a^3)^2 + 2a^3b^3 + (b^3)^2.$$

Thus, we get: $P^2 = (a^3 + b^3)^2 = (a^3)^2 + 2a^3b^3 + (b^3)^2.$

And we can put it this way, too:

$$P^2 = (a^3)^2 + 2a^3b^3 - 2a^3b^3 + 2a^3b^3 + (b^3)^2 = (a^3)^2 - 2a^3b^3 + (b^3)^2 + 4a^3b^3.$$

Thus, we get: $P^2 = (a^3)^2 - 2a^3b^3 + (b^3)^2 + 4a^3b^3.$

And we can have this, too: $(a^3 - b^3)^2 = (a^3 - b^3)(a^3 - b^3) = (a^3)^2 - 2a^3b^3 + (b^3)^2.$

So we can get: $P^2 = (a^3 - b^3)^2 + 4a^3b^3.$

Next, we have: $ab = 1$, since $a = x$, and $b = \dfrac{1}{x}$. And we have this, too: $a^3 - b^3 = 8\sqrt{5}$.

So we get: $P^2 = (a^3 - b^3)^2 + 4a^3b^3 = (8\sqrt{5})^2 + 4 = 8^2 \cdot 5 + 4 = 4(2 \cdot 8 \cdot 5 + 1) = 4 \cdot 81 = 18^2$.

Thus, we get: $P = 18$, because: $P > 0$, because: $P = x^3 + \dfrac{1}{x^3}$, and $x > 0$.

So in short:

Setting first: $a = x$, and $b = \dfrac{1}{x}$, we can get: $ab = 1$, and $a^3 - b^3 = 8\sqrt{5}$.

So next, we can get:

$$P = a^3 + b^3 \Rightarrow P^2 = (a^3 + b^3)^2 = (a^3)^2 + 2a^3b^3 + (b^3)^2 = (a^3)^2 - 2a^3b^3 + (b^3)^2 + 4a^3b^3$$

$$= (a^3 - b^3)^2 + 4a^3b^3 = (8\sqrt{5})^2 + 4 = 8^2 \cdot 5 + 4 = 4(2 \cdot 8 \cdot 5 + 1) = 4 \cdot 81 = 18^2.$$

Thus, we get: $P = 18$, because $P > 0$, since $x > 0$.

Suggestions or Solutions
To the **Problem 1** in the Example **1**

Assuming that $x^3 - \dfrac{1}{x^3} = 8\sqrt{5}$ **where** $x > 0,$ **find the values of** $Q = x + \dfrac{1}{x}.$

Setting first: $a = x$, and $b = \dfrac{1}{x}$, we can get: $ab = 1$.

And next, we can have: $(a + b)^3 = a^3 + b^3 + 3ab(a + b)$.

So we get: $Q^3 = P + 3Q$, since $P = a^3 + b^3$, and $ab = 1$. And we have: $P = 18$.

Thus, we get: $Q^3 - 3Q - P = Q^3 - 3Q - 18 = 0$.

And we can get: $Q = 3 \Rightarrow 3^3 - 3 \cdot 3 - 18 = 27 - 9 - 18 = 0$.

So we can set: $Q^3 - 3Q - 18 = (Q - 3)(uQ^2 - vQ - w)$, where u, v, and w are constant.
That is to say that $(Q - 3)$ can divide $(Q^3 - 3Q - 18)$.
So doing the synthetic division, we can get:

3	1	0	-3	-18
		3	9	18
	1	3	6	0

So we get: $(Q - 3)(Q^2 + 3Q + 6) = 0$.

$\therefore Q = 3$, because: $Q^2 + 3Q + 6 \neq 0$, because: $Q > 0$, because: $Q = x + \dfrac{1}{x}$, and $x > 0$.

And a sum of numbers positive is positive and cannot be 0.

If not quite sure of the idea behind the processes above, follow the steps below:

Setting first again: $a = x$, and $b = \dfrac{1}{x}$, we can get: $ab = 1$.

And next, we can have: $(a + b)^3 = (a + b)(a + b)^2 = (a + b)(a^2 + 2ab + b^2)$

$= a^3 + 3a^2b + 3ab^2 + b^3 = a^3 + 3ab(a + b) + b^3 = a^3 + b^3 + 3ab(a + b)$.

So we get: $(a + b)^3 = a^3 + b^3 + 3ab(a + b)$. And we have: $P = a^3 + b^3$, and $ab = 1$.

And also, we got: $P = 18$ in the example **1.0** above.   So we get: $Q^3 = P + 3Q$.

Thus, we get: $Q^3 - 3Q - P = Q^3 - 3Q - 18 = 0$.

And knowing polynomial factorizations, we can try factorizing: $Q^3 - 3Q - 18$ to solve the equation: $Q^3 - 3Q - 18 = 0$.   How?

Assuming c is constant, and $(Q - c)$ divides $(Q^3 - 3Q - 18)$, we can set:

$Q^3 - 3Q - 18 = (Q - c)(uQ^2 + vQ + w)$, where u, v, and w are constant.

Then, $Q^3 - 3Q - 18$ can be said to be factorized to $(Q - c)(uQ^2 + vQ + w)$.

Then, we can say that $Q = c$ is a solution to the equation $Q^3 - 3Q - 18 = 0$.

That's because: $Q^3 - 3Q - 18 = 0 \Rightarrow (Q - c)(uQ^2 + vQ + w) = 0$

$\Rightarrow Q - c = 0$ or $uQ^2 + vQ + w = 0$.

The value of c can be one of divisors of 18, since the equation is $Q^3 - 3Q - 18 = 0$.
So for instance, taking 3 as the value of c, we get: $Q - c = 0 \Rightarrow Q - 3 = 0 \Rightarrow Q = 3$.
Then, we get: $Q = 3 \Rightarrow Q^3 - 3Q - 18 = 3^3 - 3 \cdot 3 - 18 = 27 - 9 - 18 = 0$.

So we can set: $Q^3 - 3Q - 18 = (Q - 3)(uQ^2 - vQ - w)$, where u, v, and w are constant.
That is to say that $(Q - 3)$ can divide $(Q^3 - 3Q - 18)$.
The idea has been covered in the section **Polynomial Arithmetic 5**.
And we can easily find the quotient $(uQ^2 - vQ - w)$ using the synthetic division.

So doing the synthetic division, we can get:

$$3 \quad \bigg| \begin{array}{cccc} 1 & 0 & -3 & -18 \\ & 3 & 9 & 18 \\ \hline 1 & 3 & 6 & 0 \end{array}$$

So we get: $(Q-3)(Q^2+3Q+6)=0$.

Then, we get: $Q-3=0$ or $Q^2+3Q+6=0$.

And in this case, we get $Q=3$ only, because we get: $Q^2+3Q+6 \neq 0$.   How come?

That's because: $Q>0$, because: $Q=x+\dfrac{1}{x}$, and $x>0$, and thus, Q^2+3Q+6 is a sum of values all positive. And a sum of numbers positive is positive and cannot be 0.

And we can show: $Q^2+3Q+6 \neq 0$ the way below, too:

$$Q^2+3Q+6=Q^2+3Q+(\tfrac{3}{2})^2-(\tfrac{3}{2})^2+6=(Q+\tfrac{3}{2})^2-(\tfrac{3}{2})^2+6=(Q+\tfrac{3}{2})^2+\tfrac{15}{4} \geq \tfrac{15}{4} \text{ for all } Q.$$

- And also, we can do this problem the way below, too.

Using: $P=x^3+\dfrac{1}{x^3}=18$, and $x^3-\dfrac{1}{x^3}=8\sqrt{5}$, we can find the value of x^3, first.

$$x^3+\frac{1}{x^3}+x^3-\frac{1}{x^3}=18+8\sqrt{5} \Rightarrow 2x^3=18+8\sqrt{5} \Rightarrow x^3=9+4\sqrt{5} \Rightarrow x=\sqrt[3]{4\sqrt{5}+9}.$$

So we get: $Q=x+\dfrac{1}{x}=\sqrt[3]{4\sqrt{5}+9}+\dfrac{1}{\sqrt[3]{4\sqrt{5}+9}}$, which is equal to 3.

In addition, we can actually find the value of x directly and use the value.

That is, we can solve the equation: $x^3 - \dfrac{1}{x^3} = 8\sqrt{5}$ where $x > 0$, and use the solution.

To begin with, setting: $t = x^3$ in the equation above, we get: $t - \dfrac{1}{t} = 8\sqrt{5}$.

Thus, we get: $t - \dfrac{1}{t} = 8\sqrt{5} \Rightarrow t^2 - 1 = 8\sqrt{5}t \Rightarrow t^2 - 8\sqrt{5}t - 1 = 0$.

And using the quadratic formula, we get:

$$t = \frac{8\sqrt{5} \pm \sqrt{64 \cdot 5 + 4}}{2} = \frac{8\sqrt{5} \pm \sqrt{18^2}}{2} = \frac{8\sqrt{5} \pm 18}{2} = 4\sqrt{5} \pm 9.$$

And we have: $x > 0$, and $t = x^3$. So we get: $t > 0$. Thus, we get: $t = 4\sqrt{5} + 9$.

And next, since $t = x^3$, we get: $x = t^{\frac{1}{3}} = \sqrt[3]{t} = \sqrt[3]{4\sqrt{5} + 9}$.

₂.What is a factorization?

Normally, doing a factorization, we factorize (or factor) integers or polynomials.
Factorizing (factoring) an integer or a polynomial, we put it in terms of its *factors*, that is, we express it using its factors.
And putting it in terms of its factors, we put them in *a form of a product*.
What then, is a factor?

Factors are divisors. They are not just divisors though. What divisors then, are factors?

For instance, if we make an integer taking a product of integers, the integers we take the product of are factors, and are the factors of the integer we make.
So for instance, factorizing 12, we can put it this way: 12 = 3·4, or this way: 12 = 2·2·3.
And in the case where 12 = 3·4, we say that 3 and 4 are factors of 12, and in the case of 12 = 2·2·3, we say that the factors of 12 are two of 2s and a 3.

What then, about the cases as follows: 12 = 1·12, 12 = 1·3·4, and 12 = 1·2·2·3?

Strictly, 1 can be a factor, too, and in fact, can be a factor of every integer.
And strictly also, an integer itself can be a factor of itself, too, simply because every integer can be the product of 1 and itself.

So anyway, factorizing, we find *factors*, and put them in a *form of a product*.

Usually though, they are not just factors. Normally, if we are asked to factorize, we are to do a full factorization unless told otherwise.   What then, is a full factorization?

Doing a full factorization, we find *all the prime factors* applicable, and put in a form of a product all those prime factors. So such a factorization is called a *prime factorization* or a *prime decomposition*, too.   What do we mean by though, factors and prime factors?

Factors are divisors. Not all divisors are factors though.

A divisor can be negative as well as positive, but a factor is positive, and is a divisor. So saying a factor, we mean a positive divisor.

For instance, *all the divisors* of 12 are ±1, ±2, ±3, ±4, ±6, and ±12. So 12 has 12 divisors.

However, all the integers that can be factors of 12 are 1, 2, 3, 4, 6, and 12, each of which is a divisor of 12. It is *not* the case though, 12 has 6 factors. The biggest number of factors 12 can have is 4, since we have: $12 = 1 \cdot 2 \cdot 2 \cdot 3$. So 12 can have at most 4 factors.

So factors are divisors, but not all divisors are factors.

Talking about factorizations however, saying just a divisor, we mean a positive divisor, and for simplicity, we do not take 1 as a divisor unless told otherwise. And the same is true for a factor, too.

So we don't normally take 1 as a factor unless told otherwise or unless 1 is absolutely necessary. And thus, if an integer is said to have no factor, it can be divided by 1 and itself only. So finding such a factor cannot be an issue, and is not worth a talk.

• What's important is to find all the integers that can be factors or prime factors of an integer.

What then, is a prime factor?

A prime factor is a factor that can be divided by 1 and itself only. So for instance, if 2 is a factor, the factor is a prime factor, because it can be divided by 1 and itself only.

Usually, factorizing an integer, we find all its prime factors, and put them in a form of a product. So just saying factorizations, we mean full factorizations (prime factorizations or prime decompositions).

Normally though, we don't factorize every integer. That is, we factorize some integers. What integer then, do we factorize?

Many integers are products of integers.

More specifically, an integer can be a product of other integers.

For instance, 8 is 2·4 or 2·2·2.

What then, do we call such an integer?

It is called a *composite integer*, often just called a composite number, too.

So a composite integer is a product of other integers, and can be, for instance, 9 or 12. And most integers are composite, and thus, are composite integers.

What then, about integers not composite?

Those integers are said to be prime, and thus, are called prime integers, often just called prime numbers, too. And briefly, we just call them primes. What then, is a prime?

A prime is an integer that can be divided by 1 and itself only.

So 2 is the only prime even. And all the other primes are odd.

Among those odd primes, we have, for instance: 3, 5, 7, 11, 13, 17, 19, 23, 29, 31, etc.

So some integers called primes get combined via multiplications, and make a composite integer.

And thus, factorizing, we do a full factorization, and factorizing an integer, we factorize a composite integer, find all its prime factors, and put them in a form of a product.

So for instance, doing the full factorization of 12, we get: $12 = 2 \cdot 2 \cdot 3 = 2^2 3$, which is in a form of a product. How then, can we find those prime factors?

Putting in a form of a product, all the prime factors of an integer, what do we get?

That is, taking the product of all the prime factors of an integer, what do we get?

We get a composite integer, which is the very integer factorized.
So taking the product of all the prime factors of an integer A, for instance, we get the integer A.

And thus, we can find the prime factors doing divisions. So via divisions, we can find all those prime factors applicable, the product of which is the integer to be factorized.

And a prime factor is a prime integer, just called a prime, too.
So finding prime factors, we begin with division by the smallest prime 2, and then, move on to the division by the next larger prime, which is 3.

And thus, putting threads together, if factorizing an integer, we factorize a composite integer B, for instance. And factorizing B, we find all the prime factors applicable, that is, all the primes that can divide B, and put them in a form of a product, since the product of all those primes is the composite integer B.

So factorizing an integer, we find all the primes that divide the integer using divisions, take those divisors as factors, and paste them together using multiplication signs as · or ✕.

So for instance, factorizing 12, we can put it this way: $12 = 2 \cdot 2 \cdot 3$ or $12 = 2 ✕ 2 ✕ 3$. And those factors are prime factors, of course.

• Next, as in the case of the example above, some prime factors can be the same.

It can even be the case in fact, all prime factors are the same, too.

So for instance, we can have: $36 = 2 \cdot 2 \cdot 3 \cdot 3$, $2 \cdot 3 \cdot 2 \cdot 3$, or $3 \cdot 3 \cdot 2 \cdot 2$, and $8 = 2 \cdot 2 \cdot 2$.

In the case of 36, the prime factor 2 repeats, and so does the prime factor 3.

And in the case of 8, 2 is the only prime factor, and repeats itself.

And also, if for instance, the integer to be factorized is very big as 120932352, the factorization, that is, the product form will be too lengthy, and take up space too much.

In fact, $120932352 = 2 \cdot 2 \cdot 2 \cdot 2 \cdot 2 \cdot 2 \cdot 2 \cdot 2 \cdot 2 \cdot 2 \cdot 2 \cdot 3 \cdot 3 \cdot 3 \cdot 3 \cdot 3 \cdot 3 \cdot 3 \cdot 3 \cdot 3 \cdot 3$.

Usually therefore, we show a factorization using power notation.

So we normally put the factorizations above the way below:

$36 = 2^2 3^2$, $8 = 2^3$, and $120932352 = 2^{11} 3^{10}$.

So if an integer is fully factorized, it is expressed in terms of primes only, some of or all of which can be the same.

And thus, doing a full factorization, we find all the prime factors applicable, and put them in a form of a product. And it is often the case where it is a product of powers.

For instance, 1, 2, 3, 4, 6, 9, 12, 18, and 36 are all the integers that can be factors of 36.

And we have: $4 = 2 \cdot 2$, $6 = 2 \cdot 3$, $9 = 3 \cdot 3$, $12 = 2 \cdot 2 \cdot 3$,   and $18 = 2 \cdot 3 \cdot 3$.

So it has two kinds in prime factors, and the two are 2 and 3.

And we usually put it this way: $36 = 2^2 3^2$, and call it the (full) factorization of 36.

So in fact, we don't have to find all the divisors of the integer to be factorized.
We have only to find all its prime factors.

And note that all the prime factors of 36 are 2 of 2s and 2 of 3s, but the kinds in prime factors are 2 and 3.

Therefore, factorizing 36, we put all its prime factors in a product form, so we put together 2s and 3s the way as follows: $2 \cdot 2 \cdot 3 \cdot 3$, which is usually put this way: $2^2 3^2$.

Thus in short, factorizing 36, we get: $36 = 2^2 3^2$.

And basically, the same is true, too, for polynomial factorizations, which will begin to get covered in the next section.

3.0. Factorizing Polynomials 1

Factorizing a polynomial, too, we find its factors, and put them in a form of a product. What then, is a factor of a polynomial?

A factor is a divisor. So factorizing a polynomial, too, we find its divisors, and put them in a form of a product.   What divisor though?

Such a divisor is not just an integer but a math expression, too, which can be a constant, a monomial, or even a polynomial, too.
So a factor of a polynomial can be an *integer*, a constant, a monomial, or a polynomial.

And normally, factorizing, we do a full factorization unless told otherwise.

So factorizing a polynomial, too, we want to find all its prime factors, and put them in a form of a product.   What then, is a prime factor of a polynomial?

If an integer or an expression is prime, it has as divisors 1 and itself only.

So a prime factor of a polynomial, too, is a factor that can be divided by 1 and itself only.

Anyway, either prime or not, factors are divisors. So factorizing a polynomial, we want to find first, its divisors.   What then, can be such a divisor?

Unlike an integer factorization, such a divisor can be an integer or a math expression, which is a constant, a monomial, or a polynomial. How then, can we find such a divisor?

Factorizing a polynomial, we don't just look for a divisor that can divide the whole polynomial at once.

We look for a divisor that can divide *some of* or *all of* the terms in the polynomial. What kind of divisor is that?

We call it a *common* divisor.
If an integer or an expression divides every term in a polynomial, it divides the polynomial, and thus, is a divisor of the polynomial.

Then, such a divisor is said to be common to every term in the polynomial, and thus, is called a *common divisor*. And a common divisor can be an integer or an expression, which is a constant, a monomial, or a polynomial.

For instance, 4 is a divisor common to every term in a polynomial $4x + 8$.

That's because 4 divides $4x$, which is one term, and 4 divides 8, too, which is the other term. That is to say that 4 divides every term the polynomial $4x + 8$.

So we can say that 4 can divide the polynomial $4x + 8$. And in fact, dividing $4x + 8$ by 4, we get: $x + 2$ as the quotient, and get no remainder, that is, the remainder is 0.

And thus, we can set: $4x + 8 = 4(x + 2) = 2^2(x + 2)$, which is in a form of a product.

Normally though, we just put it this way: $4x + 8 = 4(x + 2)$.

• What then, about a polynomial $x^2 + 2x$?

We know x divides x^2, which is a term in $x^2 + 2x$, and x divides the other term, too, which is $2x$. So x is a divisor common to every term in the polynomial $x^2 + 2x$.

That is to say that x divides every term the polynomial $x^2 + 2x$.

So we can say that x can divide the polynomial $x^2 + 2x$.

And in fact, dividing $x^2 + 2x$ by x, we get: $x + 2$ as the quotient, and the remainder is 0. So we can set: $x^2 + 2x = x(x + 2)$, which is in a form of a product.

And thus, a divisor that divides every term in a polynomial is said to be common to all the terms in the polynomial, and is called a *common divisor* (of all the terms in the polynomial). And such a divisor divides the polynomial.

So a divisor common to every term in a polynomial is a divisor of the polynomial.

Normally thus, we begin with a divisor common to all the terms in the polynomial.

What then, about a factor of a polynomial?

We know a factor is a divisor.
So finding such a common divisor as stated above, we get a factor of a polynomial.

Normally though, since a factor is a divisor, we use the two terminologies 'factor' and 'divisor' interchangeably.

So using the terminology 'factor', too, we can reiterate the same story about a divisor and a common divisor described above. So let's see now, how it can get reiterated.

If an integer or an expression divides every term in a polynomial, it divides the polynomial, and thus, is a factor of the polynomial.

Then, such a *factor* is said to be *common to every term* in the polynomial, and thus, is called a *common factor* of all the terms in the polynomial. And such a common factor can be an integer or an expression, which is a constant, a monomial, or a polynomial.

For instance, x is a factor common to all the terms in a polynomial $xy + xz$.
So x divides every term in the polynomial $xy + xz$.

That is, x divides xy, and also, divides xz.

So x is a factor of xy, and is a factor of xz, too, and thus, is a factor common to both terms xy and xz. So x is a common factor of all the terms in $xy + xz$, and divides $xy + xz$.

Thus, a common factor is a common divisor, and divides the polynomial.

Then, the quotient is: $y + z$, and the remainder is 0.

And in fact, we can get: $xy + xz = x(y + z)$, which is in a form of a product.
So x is a factor of the polynomial $xy + xz$.

And the same is true for the quotient $y + z$, too, because it can divide the polynomial above. So the quotient is a factor, too, which is a polynomial, too, in this case.

And thus, a factor of a polynomial can be not only an integer, a constant, or a monomial but a polynomial, too. So factorizing a polynomial, what should we begin with?

It is a divisor or a factor common to all terms in the polynomial.
What if we cannot find it, though? Is the polynomial then, not factorable?

It is *not always* the case. What then, can we do?

Suppose for instance, we want to factorize a polynomial $xy + 2y + 3x + 6$, and call it **B**.

Then first, we can put it this way: $B = xy + 2y + 3x + 6$.

We cannot see though, any common divisor, that is, we can see no common factor. We can take however, the polynomial B as a sum of two small polynomials.

One is: $xy + 3x$, and the other is: $2y + 6$.   That is, we can put B the way below:

$B = (xy + 3x) + (2y + 6)$.

Then, we can see that each small polynomial can have a divisor, which is common to all the terms in each small polynomial, of course.

That is to say that a part of a polynomial can be another polynomial that can have a divisor.

For instance, $(xy + 3x)$ is a part of B, and x is (a divisor) common to all terms in $xy + 3x$, and thus, is a divisor of $xy + 3x$. (Note that when x is a divisor (or a factor) common to terms, we often just say that x is common to the terms.)

So we can put $(xy + 3x)$ this way: $x(y + 3)$.   What then, can we do?

We can begin with examining divisors (or factors) of each term.

And then, we can break the polynomial into parts so that all the parts can have their own divisors, and also, can share the same divisor, too.

That is, we start with examining each term, and then, partition the polynomial so that all the parts can share the same divisor.

For instance, examining each term in $B = (xy + 3x) + (2y + 6)$, we can notice that $y + 3$ is a divisor of $(xy + 3x)$, and is a divisor of $(2y + 6)$, too.   How come?

First, we can get: $xy + 3x = x(y + 3)$, since x is common to xy and $3x$.

And next, we can get: $2y + 6 = 2(y + 3)$, since 2 is common to $2y$ and 6.

And since we have: $xy + 3x = x(y + 3)$, we can say that not only x but $(y + 3)$, too, is a divisor of the polynomial $xy + 3x$, because $(y + 3)$ divides $xy + 3x$, and so does x.

And also, since we have: $2y + 6 = 2(y + 3)$, we can say that not only 2 but $(y + 3)$, too, is a divisor of the polynomial $2y + 6$, because $(y + 3)$ divides $2y + 6$, and so does 2.

So the two parts $(xy + 3x)$ and $(2y + 6)$ share the same divisor, which is: $y + 3$.
What divisor then, is the same divisor?

It is a divisor common to all the two parts $(xy + 3x)$ and $(2y + 6)$, the sum of which is B.
So what?

Taking as a term each of the two parts, we can say that the polynomial B has two terms, one is: $(xy + 3x)$, and the other is: $(2y + 6)$.

And we know: $(y + 3)$ is common to both terms $(xy + 3x)$ and $(2y + 6)$.
And we have: $B = (xy + 3x) + (2y + 6)$.

So we can say that the polynomial B has a divisor common to all the two terms, and the common divisor is: $(y + 3)$, which is thus, a factor of the polynomial B.

And thus, putting threads together, we can say that:

Setting: $B = (xy + 3x) + (2y + 6)$, and noticing x is common to every term in $(xy + 3x)$, and 2 is common to every term in $(2y + 6)$, we can see that $(y + 3)$ is a factor of B, since it is a divisor common to all the two parts $(xy + 3x)$ and $(2y + 6)$.

In other words, we can get: $B = x(y + 3) + 2(y + 3)$, where $(y + 3)$ is a common divisor.

Therefore, $(y + 3)$ divides B, and thus, is a factor of B.

So putting threads together, we get:

$B = xy + 3x + 2y + 6 = (xy + 3x) + (2y + 6) = x(y + 3) + 2(y + 3) = (y + 3)(x + 2)$.

Thus, we get: $xy + 3x + 2y + 6 = (y + 3)(x + 2)$, which is in a form of a product.

And we know: $y + 3$ has as divisors 1 and itself only, and thus, is prime. And the same is true for $x + 2$, also. So we can now say that B is (fully) factorized to $(y + 3)(x + 2)$.

• Suppose for another instance, a is a divisor common to all the terms in a polynomial P.

Then, a divides P, and thus, is a factor of P, so P can be set equal to a product of the factor a and another polynomial called Q, for instance.

That is to say that P is now set equal to aQ.   In other words, we get: $P = aQ$.

Suppose next, b is a divisor common to all the terms in the polynomial Q above.

Then, b divides Q, and thus, is a factor of Q, so Q can be set equal to a product of b and another polynomial called R, for instance.   So Q is now set equal to bR.

That is, we get: $Q = bR$.

Then, since $P = aQ$, and $Q = bR$, we get: $P = abR$.   So a, b, and R are factors of P.

Suppose now, R is: $2x^2 + 1$.

Then, no divisor is common to all the terms in R, so R has as divisors 1 and itself only.

So R is a prime factor. And in fact, a and b are prime factors, too, since each is a factor, and has as divisors 1 and itself only.

Thus, all the prime factors of P are: a, b, and R, so P is now fully factorized to abR.

In sum, we get: $P = aQ = abR = ab(2x^2 + 1)$. So before factorization, $P = 2abx^2 + ab$.

Suppose this time, R is: $2x + 6$.

Then, R can be put in $2(x + 3)$ since 2 is a divisor (or factor) common to $2x$ and 6.

Suppose also, c is: $x + 3$.   Then, we get: $R = 2c$.

And we know: $c = x + 3$ is prime, since no divisor is common to all the terms in c.

So all the prime factors of P are: 2, a, b, and c, and thus, we can say that P gets (fully) factorized to $2abc$.

In sum, we get: $P = aQ = abR = 2ab(x + 3)$. So before factorization, $P = 2abx + 6ab$.

3.1. Factorizing Polynomials 2

Factorizing a polynomial, we find its factors, and put them in a form of a product.
And a factor is a divisor. So factorizing a polynomial, we find its divisors, and put them in a form of a product.

Such a divisor can be not just an integer but a math expression, too, as a, x, ax, $ax + b$, $2x + y$, and therefore, can be a constant, a monomial, or even a polynomial, too.

So a factor of a polynomial can be an *integer*, a constant, a monomial, or a polynomial.

And normally, factorizing, we do a full factorization unless told otherwise.
So factorizing a polynomial, too, we want to find all its prime factors, and put them in a form of a product.   What then, is a prime factor of a polynomial?

If an integer or an expression is prime, it has as divisors 1 and itself only.
So a prime factor of a polynomial, too, is a factor that can be divided by 1 and itself only.
Anyway, either prime or not, factors are divisors. So factorizing a polynomial, we want to find first, its divisors.

Unlike an integer factorization, such a divisor can be an integer or a math expression, which is a constant, a monomial, or a polynomial.
And factorizing a polynomial, we don't just look for a divisor that can divide the whole polynomial at once.   What else then, can we try finding?

We can try looking for a divisor that can divide *some of* or *all of* the terms in the polynomial.   And we call it a *common* divisor.

If an integer or an expression can divide every term in a polynomial, it can divide the polynomial, and thus, is a divisor of the polynomial.

Then, such a divisor is said to be common to every term in the polynomial, and thus, is called a *common divisor*. And a common divisor can be an integer or an expression, which is a constant, a monomial, or a polynomial.

• Suppose for instance, *m* is a divisor common to some terms in a polynomial *S*, and *n* is a divisor common to all the other terms in the polynomial *S*.

Then, *S* is a sum of two products, one is a product of the divisor *m* and a polynomial, and the other is a product of the other divisor *n* and another polynomial.

Suppose next, the two polynomials in the two products above are the same, and each is called *T*.   What then, is the polynomial *T*?

The polynomial *T* is a divisor of *S*, and thus, is a factor of *S*, so *S* is the product of *T* and the sum of the two divisors *m* and *n*.   In other words, we get: $S = mT + nT = T(m + n)$.

Suppose next, *T* is a prime polynomial as $x^2 + y + 1$.

A prime polynomial is a polynomial that has as divisors 1 and itself only.

Then, *T* and $(m + n)$ are prime factors of *S*, which is thus, fully factorized to $T(m + n)$.

So we can set: $S = (m + n)(x^2 + y + 1)$, which is called the factorization of *S*.

And expanding (simplifying) the right hand side, we get: $mx^2 + my + m + nx^2 + ny + n$.

So we can say that before factorization, $S = mx^2 + my + m + nx^2 + ny + n$.

Thus, we can get: $mx^2 + my + m + nx^2 + ny + n = (m + n)(x^2 + y + 1)$.

And we can say that the polynomial S is (fully) factorized to $(m + n)(x^2 + y + 1)$.

So in a polynomial factorization, factors can be integers, constants, monomials, and polynomials that can divide the polynomial.

• Let's now, for another instance, factorize a polynomial $U = 2v^2 + v$.

First, v is a divisor common to all the terms in the polynomial U, so U can be divided by v, and therefore, we get: $U = 2v{\cdot}v + 1{\cdot}v = v(2v + 1) \Rightarrow U = v(2v + 1)$.

Next, v is a factor, and has no divisor other than 1 and itself, so v is a prime factor.

And next, the polynomial $(2v + 1)$ does not have any divisor common to all the terms in itself, but can divide U, so $(2v + 1)$ itself is a factor, and is prime, too. Thus, $(2v + 1)$ is a prime factor. Therefore, the (full) factorization of U is: $v(2v + 1)$.

• Suppose next, we want to factorize a polynomial $V = x^2 + xy + x + y$.

Then first, we try finding a divisor common to all the terms in V.

This time though, we cannot see such a divisor.   What then, should we do?

Factorizing polynomials, we basically use the **three basic laws** in arithmetic operations, which are as follows:

• Distributive Law: $A(B + C) = AB + AC$, which is in fact, a factorization.

• Associative Law: $A + B + C = (A + B) + C = A + (B + C)$, and $ABC = (AB)C = A(BC)$.

• Commutative Law: $A + B = B + A$, and $AB = BA$.

So let's now, get back to the polynomial $V = x^2 + xy + x + y$, and factorize it.

Then first, we can notice that x can divide each of the two terms in $x^2 + xy$, so using the distributive law, we can take out x from $x^2 + xy$, and put the rest in a pair of parentheses.

That is, we get: $V = x^2 + xy + x + y = x(x + y) + x + y \Rightarrow V = x(x + y) + x + y$.

Next, we can notice that $V = x(x + y) + (x + y) \cdot 1$.

Then again, assuming V is made of two terms, one is: $x(x + y)$, and the other is: $(x + y) \cdot 1$, we can see that $x + y$ can divide those two term in V. So using the distributive law, we can take out $(x + y)$, too, and put the rest in a pair of parentheses.

That is, we get: $V = x(x + y) + (x + y) \cdot 1 \Rightarrow V = (x + y)(x + 1)$.

Now, $(x + y)$ and $(x + 1)$ are factors, and are prime, too, so they are prime factors.

So the polynomial $V = x^2 + xy + x + y$ is factorized to a product of two polynomials, which are $x + y$ and $x + 1$.

Therefore, the factorization of V is: $(x + y)(x + 1)$.

And putting the factorizing processes in sum, we get:

$V = x^2 + xy + x + y = x(x + y) + x + y = (x + y)(x + 1)$.

- Let's now, try another polynomial $W = yz + xy + y^2 + xz$.

We can notice first, that y is common to xy and y^2, and can notice next, that z is common to yz and xz.   So?

So we can get: $xy + y^2 = y(z + x)$, and $yz + xz = z(y + z)$.

Then, we can see that $(z + x)$ is common to both cases.   What then?

We can factorize the polynomial the way as follows:

$$W = yz + xy + y^2 + xz = (xy + y^2) + (yz + xz) = y(z + x) + z(y + z) = (x + y)(y + z).$$

So we get: $yz + xy + y^2 + xz = (x + y)(y + z)$.

And thus, W gets factorized to $(x + y)(y + z)$.

• Let's next, try another polynomial $x^2 + 2x + 1$.

First, in $x^2 + 2x + 1$, we can notice that x can be a divisor common to two terms x^2 and $2x$. So taking x out of both of the terms, that is, dividing each of the two terms by x, we get: $x + 2$ as the sum of the quotients.

Thus, we get: $x(x + 2)$, together with 1, that is, we get: $x^2 + 2x + 1 = x(x + 2) + 1$.

However, $x + 2$ has not much to do with 1, which is the rest of the terms in the polynomial $x^2 + 2x + 1$.

That is, the sum of the quotients is not the same as, and is neither a multiple nor a divisor of the rest of the terms. So such a sum does not do the factorization any good, does it?

What should we do then?

An expression can be put a lot of ways. The same is true for a number, too. For instance:

$$10 = 5 + 5 = 3 + 7 = 12 - 2 = \frac{20}{2} = \frac{1.2}{0.12} = \dots$$

$$5x = 2x + 3x = 7x - 2x = 5x + x - x = 5x + 2x - 2x = \frac{10x}{2} = \frac{5x(y+1)}{(y+1)} = \dots$$

And doing a problem in math, we break it into parts, and then, put them together.

We don't just break it though.   We want to break it into parts so that the parts can work. That is, we break it into working parts. In this case then, such working parts should be able to give us a factor.   How then, should we break the polynomial $x^2 + 2x + 1$?

We can break it into $(x^2 + x)$ and $(x + 1)$.

Then, we get: $x^2 + 2x + 1 = (x^2 + x) + (x + 1)$.

Thus, we can get: $x(x + 1)$, together with $(x + 1)$.

That is, we get: $x^2 + 2x + 1 = x(x + 1) + x + 1$, which is $x(x + 1) + (x + 1) \cdot 1$.

So $x + 1$ is a divisor common to both of the terms, $x(x + 1)$ and $(x + 1) \cdot 1$.

Thus, we get: $x^2 + 2x + 1 = (x + 1)(x + 1) = (x + 1)^2$.

- Now, what's the opposite of a factorization?

It's an expansion. Many people call it simplification, too, though.

Usually, a polynomial is given in either of two forms: *product* or *expansion*.

That is to say that *fully* or *not*, it's been either factorized or expanded (simplified).

Suppose that a polynomial is given in a product form.

Then, it is in a product form, where expressions look pasted together by multiplications as in $\frac{2}{3} kx^2(xy + y)$ or $2x(x + 1)$, and those expressions can be polynomials, monomials, constants, or numbers.

Then, expanding or simplifying the polynomial, we remove the brackets (parentheses). We don't just erase the brackets, of course.

For instance, given a polynomial as $2(x + 3y)(x^2 + y + 1)$, which is in a product form, and is in fact, a polynomial fully factoriazed, how then, do we get the expansion?

What are we looking for when factorizing a polynomial?

It is a set of factors, which are divisors, so we mainly do divisions doing factorizations.

And expansions (simplifications) are reverse operations of factorizations.
So expanding or simplifying a polynomial factorized fully or not, we do multiplications, and thus, multiply out the factors. By the way, prime factors are often just called factors, too, for simplicity, if no confusion is involved, of course.

How then, do we do the multiplications, doing the expansion or the simplification?

We multiply out the factors.
And multiplying out factors, we multiply all the terms in all the factors in a *league*.
So for instance, expanding (simplifying) a polynomial $(x + y)(u + v + 1)$, we get:

$$(x + y)(u + v + 1) = (x + y)u + (x + y)v + (x + y) = xu + yu + xv + yv + x + y.$$

Let's for more examples, expand polynomials as follows:

$$z(xy + y)(x + z), \quad (x + y)(y + z)(z - x), \quad \text{and } (x + y + z)(x - y).$$

Then, we get:

$$z(xy + y)(x + z) = (xyz + yz)(x + z) = (xyz + yz)x + (xyz + yz)z = x^2yz + xyz + xyz^2 + yz^2.$$

$$(x + y)(y + z)(z - x) = \{x(y + z) + y(y + z)\}(z - x) = (xy + xz + y^2 + yz)(z - x)$$

$$= (xy + xz + y^2 + yz)\{z + (-x)\} = (xy + xz + y^2 + yz)z + (xy + xz + y^2 + yz)(-x)$$

$$= (xy + xz + y^2 + yz)z - (xy + xz + y^2 + yz)x = xyz + xz^2 + y^2z + yz^2 - x^2y - x^2z - y^2x - xyz$$

$$= xz^2 + y^2z + yz^2 - x^2y - x^2z - y^2x.$$

$$(x + y + z)(x - y) = (x + y + z)x - (x + y + z)y = x^2 + xy + xz - (xy + y^2 + yz)$$

$$= x^2 + xy + xz - xy - y^2 - yz = x^2 + xz - y^2 - yz.$$

And doing the operations above backward, we can see better how factorizations can go.

Now, factorizations of some polynomials are frequently used when we do algebra. They are called *factorization identities*, which are often called factorization formulas, too. And the factorization identities often used are as follows:

$$(x + y)^2 = x^2 + 2xy + y^2. \qquad (x + y)^3 = x^3 + 3x^2y + 3xy^2 + y^3.$$

$$(x + y)(x - y) = x^2 - y^2. \qquad (x + y)(x^2 - xy + y^2) = x^3 + y^3.$$

$$(x^2 + xy + y^2)(x^2 - xy + y^2) = x^4 + x^2y^2 + y^4.$$

$$(x + a)(x + b) = x^2 + (a + b)x + ab. \qquad (ax + b)(cx + d) = acx^2 + (ad + bc)x + bd.$$

$$(x + a)(x + b)(x + c) = x^3 + (a + b + c)x^2 + (ac + bc + ca)x + abc.$$

$$(a + b + c)^2 = a^2 + b^2 + c^2 + 2(ab + bc + ca).$$

$$(a + b + c)(a^2 + b^2 + c^2 - ab - bc - ca) = a^3 + b^3 + c^3 - 3abc.$$

And of course, we are going to see how we can get the left hand side from the right hand side of each of all the identities above when doing the sets of **Examples**.
And doing some algebra to some of those above, we can get some more identities.

Changing y with **-y**, we can get:

$$(x + y)^2 = x^2 + 2xy + y^2 \Rightarrow (x - y)^2 = x^2 - 2xy + y^2.$$

$$(x + y)^3 = x^3 + 3x^2y + 3xy^2 + y^3 \Rightarrow (x - y)^3 = x^3 - 3x^2y + 3xy^2 - y^3.$$

$$(x + y)(x^2 - xy + y^2) = x^3 + y^3 \Rightarrow (x - y)(x^2 + xy + y^2) = x^3 - y^3.$$

And by the same token, we can get:

$(x + a)(x + b) = x^2 + (a + b)x + ab \Rightarrow (x - a)(x - b) = x^2 - (a + b)x + ab.$

$(ax + b)(cx + d) = acx^2 + (ad + bc)x + bd \Rightarrow (ax - b)(cx - d) = acx^2 - (ad + bc)x + bd.$

$(x + a)(x + b)(x + c) = x^3 + (a + b + c)x^2 + (ac + bc + ca)x + abc \Rightarrow$

$(x - a)(x - b)(x - c) = x^3 - (a + b + c)x^2 + (ac + bc + ca)x - abc.$

And we can put together some of those above the way below:

To begin with, we have: $(x + y)^2 = x^2 + 2xy + y^2$, and $(x - y)^2 = x^2 - 2xy + y^2$.

So putting both together, we can have: $(x \pm y)^2 = x^2 \pm 2xy + y^2$.

Next, we have: $(x + y)^3 = x^3 + 3x^2y + 3xy^2 + y^3$, and $(x - y)^3 = x^3 - 3x^2y + 3xy^2 - y^3$.

So putting both together, we can have: $(x \pm y)^3 = x^3 \pm 3x^2y + 3xy^2 \pm y^3$.

And we can have:

$x^3 + 3x^2y + 3xy^2 + y^3 = x^3 + 3xy(x + y) + y^3$ & $x^3 - 3x^2y + 3xy^2 - y^3 = x^3 - 3xy(x - y) - y^3$.

So we can put it this way, too: $x^3 \pm 3x^2y + 3xy^2 \pm y^3 = x^3 \pm 3xy(x \pm y) \pm y^3$.

Next, we have: $(x + y)(x^2 - xy + y^2) = x^3 + y^3$, and $(x - y)(x^2 + xy + y^2) = x^3 - y^3$.

So putting both together, we can have: $x^3 \pm y^3 = (x \pm y)(x^2 \mp xy + y^2)$.

Next, we have: $(x + a)(x + b) = x^2 + (a + b)x + ab$, & $(x - a)(x - b) = x^2 - (a + b)x + ab$.

So putting both together, we can have: $(x \pm a)(x \pm b) = x^2 \pm (a + b)x + ab$.

And next, we have:

$$(x + a)(x + b)(x + c) = x^3 + (a + b + c)x^2 + (ac + bc + ca)x + abc.$$

$$(x - a)(x - b)(x - c) = x^3 - (a + b + c)x^2 + (ac + bc + ca)x - abc.$$

So in sum, we have: $(x \pm a)(x \pm b)(x \pm c) = x^3 \pm (a + b + c)x^2 + (ac + bc + ca)x \pm abc.$

And also, doing some algebra, we can get some useful expressions as follows:

$$(x + y)^2 = x^2 + 2xy + y^2 \Rightarrow x^2 + y^2 = (x + y)^2 - 2xy.$$

$$(x - y)^2 = x^2 - 2xy + y^2 \Rightarrow x^2 + y^2 = (x - y)^2 + 2xy.$$

$$(x + y)^3 = x^3 + 3x^2y + 3xy^2 + y^3 = x^3 + 3xy(x + y) + y^3 \Rightarrow x^3 + y^3 = (x + y)^3 - 3xy(x + y).$$

$$(x - y)^3 = x^3 - 3x^2y + 3xy^2 - y^3 = x^3 - 3xy(x + y) \pm y^3 \Rightarrow x^3 - y^3 = (x - y)^3 + 3xy(x + y).$$

And also, we can put $a^3 + b^3 + c^3 - 3abc = (a + b + c)(a^2 + b^2 + c^2 - ab - bc - ca)$ this way, too: $a^3 + b^3 + c^3 - 3abc = \frac{1}{2}(a + b + c)\{(a - b)^2 + (b - c)^2 + (c - a)^2\}.$

Examples 1

Factorize (that is, factor) the polynomials below:

0. $2a + 4b + 16c$

1. $2ab + 6bc$

2. $2ac + 6bc + ad + 3bd$

3. $3k^2(b + 2b^2c) + 9bk + 27b^2k^4$

Suggestions or Solutions
To the **Problem** in the Example **0**

This example is just a warming-up, and we have: $2a + 4b + 16c$, which is a polynomial.

And doing this example, we are just going over the basics in polynomial factorization processes.

So to begin with, what are we looking for when factorizing a polynomial?

We are looking for factors, which divide the polynomial.
So we want to find a divisor. Finding it though, we don't just find a divisor that divides the whole polynomial at once. What else then, do we do?

Normally, we begin with a divisor that can divide every term the polynomial has.
And such a divisor is said to be common to all the terms in the polynomial, and thus, is called a common divisor.

So a common divisor can divide every term in the polynomial, and thus, divides the polynomial, and is a factor of the polynomial. So what does a factor do?

A factor is a divisor, and a factor of a polynomial divides the polynomial.
A divisor common to all the terms in a polynomial is a factor of the polynomial.

Suppose now, $P = 2a + 4b + 16c$.

Then, examining all terms in the polynomial P, we can see 2 is a divisor common to all the terms, and thus, 2 is a factor of P.

We have: $2a = 2 \cdot a$, $4b = 2 \cdot 2b$, and $16c = 2 \cdot 8c$, so **2** is a common divisor.   So what?

So factorizing the polynomial P, we divide by 2 every term in P, put every quotient in a pair of parentheses, and take a product of the divisor 2 and the *sum* of all the quotients.

Then, we get: $P = 2a + 4b + 16c = 2(a + 2b + 8c)$, where $a + 2b + 8c$ is the *sum* stated above.

So we can see that 2 divides the polynomial P, and so does the sum of those quotients.

Therefore, 2 and the sum are factors of P.

What then, about the sum, which is a polynomial, too?

If an integer or an expression as a polynomial has as divisors 1 and itself only, it is said to be prime. And if a factor is prime, the factor is called a prime factor.

Now, the sum is the polynomial $a + 2b + 8c$, and no divisor is common to all terms in the polynomial. That is, $a + 2b + 8c$ has no divisor other than 1 and itself.

So $a + 2b + 8c$ is prime, is a factor of P, and thus, is a prime factor of P.
Besides, 2 is a prime factor of P, too.

And factorizing a polynomial, we find all its prime factors, and put them in a form of a product.

Thus, the polynomial $P = 2a + 4b + 16c$ gets (fully) factorized to $2(a + 2b + 8c)$, which is in a form of a product.

In short:

$2a + 4b + 16c = 2(a + 2b + 8c).$

Suggestions or Solutions
To the **Problem** in the Example **1**

We have: $2ab + 6bc$.

Factorizing it, we get: $2ab + 6bc = 2b(a + 3c)$.

If not quite sure of how we can get it, follow the steps below:

To begin with, what are we looking for when doing factorizations?

We want to find divisors, which are factors.

What then, can be a divisor, that is, a factor when we factorize a polynomial?

Of course, we want to find a divisor that divides the polynomial, and such a divisor is a factor of the polynomial. What then, can be such a divisor?

It can be an integer or an expression as a constant, monomial, or a polynomial.

And normally, we begin with a divisor that can divide every term the polynomial has. And such a divisor is called a common divisor.

So a common divisor can divide every term in the polynomial, and thus, divides the polynomial, and is a factor of the polynomial. What then, can be such a common divisor? That is, what divides every term in the polynomial given in this problem?

We can see 2 does, because it is a divisor common to all the terms. Is that it though?

So does **b**. Thus, 2 and **b** both can divide every term in the polynomial, so **2b** divides the polynomial, too. That is to say that **2b** is a divisor common to all the terms in the polynomial, and thus, is a factor of the polynomial.

So we want to divide by **2b** every term in the polynomial, put every quotient in a pair of parentheses, and then, take a product of the divisor and the sum of all the quotients.

Suppose now, $P = 2ab + 6bc$.

Then, we get: $P = 2ab + 6bc = 2{\cdot}ab + 2{\cdot}3bc = 2b{\cdot}a + 2b{\cdot}3c = 2b(a + 3c)$.

So we can see that the polynomial $a + 3c$ is a factor, too, of the polynomial P, of course.

What factor then, is $a + 3c$?   In other words, is the factor prime?

There is no divisor common to all terms in the polynomial $a + 3c$.
So the polynomial $a + 3c$ is a prime factor of P. And so are **2** and **b**.

So all the prime factors found are an integer **2**, a monomial **b**, and a polynomial $a + 3c$.

And factorizing a polynomial, we put all its prime factors in a form of a product.
So the factorization of the polynomial $P = 2ab + 6bc$ is: $2b(a + 3c)$.

Then, we can say that the polynomial P gets fully factorized to $2b(a + 3c)$.

In short:

$2ab + 6bc = 2b(a + 3c)$.

By the way, we don't normally factorize a monomial.
We can factorize it though if it has an integer factorable. And if it does, and we are asked to factorize it, we need to factorize it.
So for instance, if we really need to factorize **8b**, and factorizing it, we get: 2^3b.

Suggestions or Solutions
To the Problem in the Example 2

We have: $2ac + 6bc + ad + 3bd$.   And factorizing it, we get:

$$2ac + 6bc + ad + 3bd = 2c(a + 3b) + d(a + 3b) = (a + 3b)(2c + d).$$

If not quite sure of the idea behind the processes above, follow the steps below:

To begin with, it's a good idea to name a polynomial.

So let's suppose first, $P = 2ac + 6bc + ad + 3bd$.

Unlike the previous examples, we do not see this time, a divisor common to all the terms in the polynomial P.
So can we say that the polynomial P itself is prime, and thus, is not factorable?

Though a polynomial doesn't seem to have a divisor common to all its terms, it can still be factorable.

A polynomial can be a sum of smaller polynomials, and each of those smaller polynomials can have a divisor common to all the terms in itself.

That is to say that there can be divisors, each of which can divide each of the smaller polynomials, the sum of which is of course, the polynomial that is to be factorized.

In addition, after all such divisions, all the quotients can be the same.
What then, can the same quotient do?

The same quotient can divide the polynomial, and thus, is a factor of the polynomial.

That is to say that the polynomial is the product of the same quotient and the sum of the divisors stated above.   What then, do we call the sum?

The sum is called a factor of the polynomial, too, because it can divide the polynomial, too. So the polynomial can be put in a form of a product of two, one is the same quotient, and the other is the sum.

So let's now, take a closer look at the polynomial P, and see if it can be portioned so that we can extract factors the way described above.

Now, examining all the terms in P, we can see that $2c$ is a divisor common to two terms, $2ac$ and $6bc$, and that d is a divisor common to the other two terms, ad and $3bd$.

So $2c$ is a divisor of a polynomial $2ac + 6bc$, and d divides a polynomial $ad + 3bd$.

Thus, we get:

$$P = 2ac + 6bc + ad + 3bd = (2c \cdot a + 2c \cdot 3b) + (a \cdot d + 3b \cdot d) = 2c(a + 3b) + d(a + 3b).$$

And thus, after all the divisions, we get the same quotient, which is: $a + 3b$, which therefore, is a divisor common to both of $2c(a + 3b)$ and $d(a + 3b)$.

So $a + 3b$ can divide P, and thus, is a factor of P.
Thus, dividing P by $a + 3b$, what do we get?

We get: $2c + d$, which is the sum of the two divisors $2c$ and d.

In other words, $P = 2c(a + 3b) + d(a + 3b) = (a + 3b)(2c + d) \Rightarrow P = (a + 3b)(2c + d)$.

So we can see that $2c + d$, too, is a divisor of P, and is a factor of P.

Now, $(a + 3b)$ and $(2c + d)$ both are prime, and are all the factors of P, so they are all the prime factors of P, and therefore, the factorization of P is now complete.

In short:

$$2ac + 6bc + ad + 3bd = 2c(a + 3b) + d(a + 3b) = (a + 3b)(2c + d).$$

Suggestions or Solutions
To the **Problem** in the Example **3**

We have: $3k^2(b + 2b^2c) + 9bk + 27b^2k^4$.

Factorizing it, we get:

$3k^2(b + 2b^2c) + 9bk + 27b^2k^4$

$= k\{3k(b + 2b^2c) + 9b + 27b^2k^3\}$

$= k\{3kb(1 + 2bc) + 9b(1 + 3bk^3)\}$

$= 3kb\{k(1 + 2bc) + 3(1 + 3bk^3)\}$

$= 3bk(k + 2bck + 3 + 9bk^3)$.

If not quite sure of the idea behind the processes above, follow the steps below:

Suppose $P = 3k^2(b + 2b^2c) + 9bk + 27b^2k^4$.

To begin with, we can see k is a divisor common to all the terms, so k is a factor of P.

Thus, we can take out k from every term in P, since k is common to all the terms in P.

Then, we get: $P = k\{3k(b + 2b^2c) + 9b + 27b^2k^3\}$, for now.

Next, we can notice that b is a divisor common to all the terms, too, so b is a factor of P, also. How come?

Suppose Q is the polynomial inside the curly brackets above.

Then, cutting Q into two parts, we can set: $Q = 3k(b + 2b^2c) + (9b + 27b^2k^3)$.

So we get two parts, which are: $3k(b + 2b^2c)$ and $9b + 27b^2k^3$.

Then, we can see that b is a divisor common to $b + 2b^2c$, and also, is a divisor common to $9b + 27b^2k^3$, too.

Thus, we get: $b + 2b^2c = b(1 + 2bc)$, and $9b + 27b^2k^3 = b(9 + 27bk^3)$.

So we get: $Q = 3k(b + 2b^2c) + (9b + 27b^2k^3) = 3kb(1 + 2bc) + b(9 + 27bk^3)$.

Thus, we can see that b is a divisor common to $3kb(1 + 2bc)$ and $b(9 + 27bk^3)$.

So we get: $Q = 3kb(1 + 2bc) + b(9 + 27bk^3) = b\{3k(1 + 2bc) + (9 + 27bk^3)\}$.

Thus, we get: $Q = 3k(b + 2b^2c) + 9b + 27b^2k^3 = b\{3k(1 + 2bc) + (9 + 27bk^3)\}$.

Now, we have: $P = kQ$.

So we get: $P = k\{3k(b + 2b^2c) + 9b + 27b^2k^3\} = kb\{3k(1 + 2bc) + (9 + 27bk^3)\}$.

Therefore, b is a factor of P, too.   Then, are we there now?

Not quite. We have another factor of P, which is 3.   How come?

We can see that $(9 + 27bk^3)$ has a divisor, which is **9**.

So we get: $P = kb\{3k(1 + 2bc) + (9 + 27bk^3)\} = kb\{3k(1 + 2bc) + 9(1 + 3bk^3)\}$.

And in turn, we can see 3 is a common divisor in what's inside the curly brackets above.

What's inside the brackets can be taken as a polynomial with two terms, and 3 is common to the two.   What then, are the two?

One is: $3k(1 + 2bc)$, and the other is: $9(1 + 3bk^3)$.

So we can take it out of each of the two terms, and then, put it outside the brackets.

Then, we get:

$$P = kb\{3k(1 + 2bc) + 9(1 + 3bk^3)\} = kb[3\{k(1 + 2bc) + 3(1 + 3bk^3)\}]$$

$$= 3kb\{k(1 + 2bc) + 3(1 + 3bk^3)\}. \quad \text{Are we there then?}$$

Almost.

We want to expand the two terms inside the brackets, and then, check to see if we can proceed further with the factorization. So expanding the two first, we get:

$$P = 3kb\{k(1 + 2bc) + 3(1 + 3bk^3)\} = 3bk(k + 2bck + 3 + 9bk^3).$$

And next, we want to check to see if what's inside the parentheses can be factorized. What's inside the parentheses is a polynomial, and we can put it the way below:

$k + 2bck + 3 + 9bk^3 = k(1 + 2bc) + 3(1 + 3bk^3)$, which has however, no divisor common to $k(1 + 2bc)$ and $3(1 + 3bk^3)$.

We can put it the way below, too:

$k + 2bck + 3 + 9bk^3 = k + 2bck + 9bk^3 + 3 = k(1 + 2bc + 9bk^2) + 3$, which however, has no divisor common to $k(1 + 2bc + 9bk^2)$ and 3.

We can put it this way, too: $k(1 + 2bc + 9bk^2) + 3 = k\{1 + b(2c + 9k^2)\} + 3$, which however, does not give us any divisor.

We can put it the way below, too:

$k + 2bck + 3 + 9bk^3 = k + 3 + 2bck + 9bk^3 = k + 3 + bk(2c + 9k^2)$, which however, has no divisor common to k, 3, and $bk(2c + 9k^2)$.

So we can conclude that the polynomial $k + 2bck + 3 + 9bk^3$ has no divisor other than 1 and itself, and therefore, is prime.

Thus, the factorization can now be completed as follows:

$$P = 3bk(k + 2bck + 3 + 9bk^3).$$

And all the prime factors of P are: 3, b, k, and $k + 2bck + 3 + 9bk^3$.

In short:

$$3k^2(b + 2b^2c) + 9bk + 27b^2k^4$$

$$= k\{3k(b + 2b^2c) + 9b + 27b^2k^3\}$$

$$= k\{3kb(1 + 2bc) + 9b(1 + 3bk^3)\}$$

$$= 3kb\{k(1 + 2bc) + 3(1 + 3bk^3)\}$$

$$= 3bk(k + 2bck + 3 + 9bk^3).$$

And of course, there can be other ways we can get the factorization, too.

Examples 2

0. Factorize the polynomials below.

0.0. $x^2 + 2xy + y^2$

0.1. $x^2 - 2xy + y^2$

0.2. $x^3 + 3x^2y + 3xy^2 + y^3$

0.3. $x^3 - 3x^2y + 3xy^2 - y^3$

1. Assuming that $a^2 + b^2 + 2ab = K^2$, put K in terms of a and b.

Suggestions or Solutions
To the **Problem 0** in the Example **0**

We have $x^2 + 2xy + y^2$.

Factorizing it, we get:

$$x^2 + 2xy + y^2 = x^2 + xy + xy + y^2 = x(x + y) + y(x + y) = (x + y)(x + y) = (x + y)^2.$$

If not quite sure of the idea behind the processes above, follow the steps below:

Doing math, we often need to do mental math, together with putting down numbers. So working in math, we don't just work with numbers doing arithmetic and algebra but keep running a stream of logic, too, along with math operations and activities.

Doing mental math, we work on particular numbers or expressions taking care of the next operations keeping in mind results of the current operations. Besides, we do other activities as substitutions.

And it is particularly the case when we do factorizations. Quite often, we need to make some predictions, too, doing factorizations. Such predictions are not random guesses but logical expectations, which are in fact, what mental math is mostly about.

Now, factorizing a polynomial, we find factors of the polynomial, and put them in a form of a product.   How then, can we find the factors?

Factors are divisors. So when factorizing a polynomial, we look for divisors, each of which can divide the polynomial.   How then, can we find such divisors?

Normally, when factorizing a polynomial, we try first, finding a divisor that can divide each and every term in the polynomial. Such a divisor is called a common divisor, and does divide the polynomial, since it divides every term in the polynomial.

So if an integer or an expression divides every term in a polynomial, it is said to be common to all the terms in the polynomial, and is called a common divisor.

What then, does the integer or the expression do?

It divides the polynomial, and thus, is a divisor of the polynomial. And we call such a divisor a factor of the polynomial.

It's not always the case however, we find a factor the way above.
In other words, that's not the only way we can find a factor of a polynomial.
The principle remains the same though.
That is to say that we often need to look at the same problem in a different perspective.

Suppose for instance, we can find a divisor that divides not every term but some terms.

Then, it can be the case where we get a factor, which divides the polynomial, of course. How can we get such a factor though?

Doing a problem in math, we break it apart, and put the pieces together.
We don't just break it though.

We want to break it apart so that we can form the solution putting the pieces together. And we want to keep in mind that the principle remains the same.

Now, setting: $P = x^2 + 2xy + y^2$, we can say that P doesn't look factorable, because we don't see any divisor common to all its terms.   What then, can we do?

A polynomial can be a sum of polynomials.

So taking apart $2xy$ into two parts, we can put P the way as follows:

$P = x^2 + xy + xy + y^2 = (x^2 + xy) + (xy + y^2)$, which can be thus, taken as a polynomial made of two terms, one is: $(x^2 + xy)$, and the other is: $(xy + y^2)$. So what?

It was mentioned that the principle remains the same. What then, is the principle?

• Finding a factor of a polynomial, we find a divisor common to all the terms.

So looking closely at each of the two terms $(x^2 + xy)$ and $(xy + y^2)$, we can see x is common to all the terms in $x^2 + xy$, and y is common in $xy + y^2$.

So we get: $P = x^2 + xy + xy + y^2 = x(x + y) + y(x + y)$, where $x + y$ is a divisor common to both terms, $x(x + y)$ and $y(x + y)$.

So we get: $x(x + y) + y(x + y) = (x + y)(x + y)$, and thus, we get: $P = (x + y)^2$.

Therefore, $x^2 + 2xy + y^2$ is factorized to $(x + y)^2$.

In short:

$x^2 + 2xy + y^2 = x^2 + xy + xy + y^2 = x(x + y) + y(x + y) = (x + y)(x + y) = (x + y)^2$.

Suggestions or Solutions
To the **Problem 1** in the Example **0**

We have: $x^2 - 2xy + y^2$.

This is in fact, no other than the previous problem.
We can notice that putting $-y$ into y in $x^2 + 2xy + y^2$, we get: $x^2 - 2xy + y^2$.
And we have: $x^2 + 2xy + y^2 = (x + y)^2$.

So in $(x + y)^2$, replacing y with $-y$, we can readily get: $(x - y)^2$, which is P factorized.

For practice purposes though, let's get the solution the way we get the previous one.

It's a good idea to name a polynomial if it's not named. So set first: $P = x^2 - 2xy + y^2$.

And next, taking apart $2xy$ into two terms, we get: $P = x^2 - 2xy + y^2 = x^2 - xy - xy + y^2$.
That is, we get:
$P = x^2 - 2xy + y^2 = x^2 - xy - xy + y^2 = (x^2 - xy) - (xy - y^2)$, which can be thus, taken as a polynomial made of two terms, one is: $(x^2 - xy)$, and the other is: $(xy - y^2)$.

Then, we can see x is common in $x^2 - xy$, and y is common in $xy - y^2$.

So we get: $P = (x^2 - xy) - (xy - y^2) = x(x - y) - y(x - y)$, where $x - y$ is a divisor common to both terms, $x(x - y)$ and $y(x - y)$.

Thus, we get: $P = x(x - y) - y(x - y) = (x - y)(x - y)$, which is $(x - y)^2$.
Therefore, we get: $P = (x - y)^2$.

In short:

$x^2 - 2xy + y^2 = x^2 - xy - xy + y^2 = x(x - y) + y(-x + y)$

$= x(x - y) - y(x - y) = (x - y)(x - y) = (x - y)^2$.

Suggestions or Solutions
To the **Problem 2** In the Example **0**

We have: $x^3 + 3x^2y + 3xy^2 + y^3$.

Factorizing it, we can begin with:

$x^3 + 3x^2y + 3xy^2 + y^3 = (x^3 + 3x^2y) + (3xy^2 + y^3) = x(x^2 + 3xy) + (3xy + y^2)y$.

So first, $x^2 + 3xy = x^2 + 2xy + xy + y^2 - y^2 = x^2 + 2xy + y^2 + xy - y^2 = (x + y)^2 + xy - y^2$.

Thus, we get: $x(x^2 + 3xy) = x(x + y)^2 + x^2y - xy^2$.

Next, $3xy + y^2 = x^2 - x^2 + 2xy + xy + y^2 = x^2 + 2xy + y^2 + xy - x^2 = (x + y)^2 + xy - x^2$.

So we get: $(3xy + y^2)y = y(x + y)^2 + xy^2 - x^2y$.

Thus, we get:

$x(x^2 + 3xy) + (3xy + y^2)y = \{x(x + y)^2 + x^2y - xy^2\} + \{y(x + y)^2 + xy^2 - x^2y\}$

$= x(x + y)^2 + y(x + y)^2 + x^2y - xy^2 + xy^2 - x^2y = x(x + y)^2 + y(x + y)^2$

$= (x + y)^2(x + y) = (x + y)^3$.

If not quite sure of the idea behind the processes above, follow the steps below:

Setting first, $P = x^3 + 3x^2y + 3xy^2 + y^3$, we can say that P has a symmetry.

Even if x and y get exchanged, we get exactly the same polynomial P. So we should be able to take advantage of it.

Thus next, we can try breaking the polynomial P into two parts the way below:

$P = x^3 + 3x^2y + 3xy^2 + y^3 = (x^3 + 3x^2y) + (3xy^2 + y^3)$.

Then, we can see a couple of cases where we can get a divisor common to all the terms in each part.

In one case, x is common in the first part, and y is common in the second part.

So we can put the polynomial P the way below:

$$P = (x^3 + 3x^2y) + (3xy^2 + y^3) = x(x^2 + 3xy) + (3xy + y^2)y.$$

And next, getting back to polynomial $(x^3 + 3x^2y) + (3xy^2 + y^3)$, we can see the other case, where x^2 is common in the first part, and y^2 is common in the second part.

So this time, we can put the polynomial P the way below:

$$P = (x^3 + 3x^2y) + (3xy^2 + y^3) = x^2(x + 3y) + (3x + y)y^2.$$

Which case then, do we want to take?

We want to take the first case, where $P = x(x^2 + 3xy) + (3xy + y^2)y$.

That's because we can expect the polynomial $x^2 + 3xy$ to be converted to '$x^2 + 2xy + y^2$' seeing that $x^2 + 3xy$ has $x^2 + 2xy$, so we can get: $(x + y)^2$ if we add y^2 to it.

So let's first, take care of $x(x^2 + 3xy)$ keeping in mind that we have: $(3xy + y^2)y$, too.

We don't overpay, underpay, or compromise doing math. In math, we can compensate only. That is, getting one thing in math, we give the one exactly the same or exactly equivalent. So math is always a fair business, isn't it? It is fair and square, of course, and is always so.

Now, setting: $3xy = 2xy + xy$, and then, adding and subtracting y^2, we can make $(x + y)^2$.

In other words, we can get:

$$x^2 + 3xy = x^2 + 2xy + xy + y^2 - y^2 = \underline{x^2 + 2xy + y^2} + xy - y^2 = \underline{(x + y)^2} + xy - y^2.$$

So we get: $x(x^2 + 3xy) = x\{(x + y)^2 + xy - y^2\} = x(x + y)^2 + x^2y - xy^2.$

Let's now, move on to the other part, which is: $(3xy + y^2)y$.
What then, can we do about it?

We know the polynomial P is symmetric.　So?

So doing to the other part the same thing as the one we did to the part $x(x^2 + 3xy)$, because of the symmetry, we get the same result. That is, we get:

$$3xy + y^2 = x^2 - x^2 + 2xy + xy + y^2 = \underline{x^2 + 2xy + y^2} + xy - x^2 = \underline{(x + y)^2} + xy - x^2.$$

So we get: $(3xy + y^2)y = y(x + y)^2 + xy^2 - x^2y.$

And thus, putting threads together, we get:

$$P = x(x^2 + 3xy) + (3xy + y^2)y = \{x(x + y)^2 + x^2y - xy^2\} + \{y(x + y)^2 + xy^2 - x^2y\}$$

$$= x(x + y)^2 + y(x + y)^2 + x^2y - xy^2 + xy^2 - x^2y = x(x + y)^2 + y(x + y)^2.$$

Then, we can see that $(x + y)^2$ is a divisor common to $x(x + y)^2$ and $y(x + y)^2$.

So $(x + y)^2$ is a divisor of P, and thus, can be a factor of P.

That is to say that we get: $P = x(x + y)^2 + y(x + y)^2 = (x + y)^2(x + y) = (x + y)^3$, which is the factorization of P.

In short:

$$x^3 + 3x^2y + 3xy^2 + y^3 = (x^3 + 3x^2y) + (3xy^2 + y^3) = x(x^2 + 3xy) + (3xy + y^2)y.$$

So first, $x^2 + 3xy = x^2 + 2xy + xy + y^2 - y^2 = x^2 + 2xy + y^2 + xy - y^2 = (x + y)^2 + xy - y^2.$

Thus, we get: $x(x^2 + 3xy) = x(x + y)^2 + x^2y - xy^2.$

Next, $3xy + y^2 = x^2 - x^2 + 2xy + xy + y^2 = x^2 + 2xy + y^2 + xy - x^2 = (x + y)^2 + xy - x^2.$

So we get: $(3xy + y^2)y = y(x + y)^2 + xy^2 - x^2y.$

Thus, we get:

$$x(x^2 + 3xy) + (3xy + y^2)y = \{x(x + y)^2 + x^2y - xy^2\} + \{y(x + y)^2 + xy^2 - x^2y\}$$

$$= x(x + y)^2 + y(x + y)^2 + x^2y - xy^2 + xy^2 - x^2y = x(x + y)^2 + y(x + y)^2$$

$$= (x + y)^2(x + y) = (x + y)^3.$$

And we can have a sequence as follows:

$$x + y = x + y,$$

$$(x + y)^2 = x^2 + 2xy + y^2,$$

$$(x + y)^3 = x^3 + 3x^2y + 3xy^2 + y^3,$$

$$(x + y)^4 = x^4 + 4x^3y + 6x^2y^2 + 4xy^3 + y^4,$$

$$(x + y)^5 = x^5 + 5x^4y + 10x^3y^2 + 10x^2y^3 + 5xy^4 + y^5,$$

...

What then, are the expansions of $(x + y)^{1000}$ and $(x + y)^k$, where k is a nonnegative integer?

Suggestions or Solutions
To the Problem 3 in the Example 0

We have: $x^3 - 3x^2y + 3xy^2 - y^3$.

Factorizing it, we can begin with:

$x^3 - 3x^2y + 3xy^2 - y^3 = (x^3 - 3x^2y) + (3xy^2 - y^3) = x(x^2 - 3xy) - (-3xy + y^2)y$.

First, $x^2 - 3xy = x^2 - 2xy - xy + y^2 - y^2 = x^2 - 2xy + y^2 - xy - y^2 = (x - y)^2 - xy - y^2$.

So we get: $x(x^2 - 3xy) = x(x - y)^2 - x^2y - xy^2$.

Next, $-3xy + y^2 = x^2 - x^2 - 2xy - xy + y^2 = x^2 - 2xy + y^2 - xy - x^2 = (x - y)^2 - xy - x^2$.

So we get: $-y(-3xy + y^2) = -y(x - y)^2 + xy^2 + x^2y$.

Thus, we get:

$x(x^2 - 3xy) - y(-3xy + y^2) = x(x^2 - 3xy) + \{-y(-3xy + y^2)\}$

$= \{x(x - y)^2 - x^2y - xy^2\} + \{-y(x - y)^2 + xy^2 + x^2y\}$

$= x(x - y)^2 - y(x - y)^2 - x^2y - xy^2 + xy^2 + x^2y = x(x - y)^2 - y(x - y)^2$

$= (x - y)^2(x - y) = (x - y)^3$.

If not quite sure of the idea behind the processes above, follow the steps below:

Suppose first, $P = x^3 - 3x^2y + 3xy^2 - y^3$.
The polynomial P is quite similar to the one in the problem 2 above.
In fact, P is as good as the one in the problem 2. How come?

It's just a matter of changing y with $-y$.
Substituting y with $-y$ in P, we get the one in the problem 2.
So putting $-y$ into y in $x^3 - 3x^2y + 3xy^2 - y^3$, we get:

$x^3 - 3x^2(-y) + 3xy^2 - (-y)^3 = x^3 + 3x^2y + 3xy^2 + y^3$.

And we know: $x^3 + 3x^2y + 3xy^2 + y^3 = (x + y)^3$.

So changing y with $-y$ in the polynomial in the problem 2, we get:

$x^3 - 3x^2y + 3xy^2 - y^3 = (x - y)^3$.

Let's see though, how the factorization of P can actually proceed in a usual manner.

There is a symmetry in P: changing x with $-y$, and y with $-x$. we get the same P.
So we should take advantage of it.

And the factorization method or procedure will be just about the same as the one we used for the previous example. So first, we break P into two parts.
And next, we take one part at a time keeping in mind the other part to be taken care of.

So to begin with, x is common in $x^3 - 3x^2y$, so we get: $x^3 - 3x^2y = x(x^2 - 3xy)$.

Next, $-y$ is common in $3xy^2 - y^3$, so we get: $3xy^2 - y^3 = -y(-3xy + y^2)$.
Why is $-y$ common though?

We can get: $3xy^2 - y^3 = 3x(-y)(-y) + (-y^3) = 3x(-y)(-y) + (-y)y^2$, where $-y$ is common.

Now, beginning with $x^2 - 3xy$, we can first, put it the way below:

$x^2 - 3xy = x^2 - 2xy - xy + y^2 - y^2 = \underline{x^2 - 2xy + y^2} - xy - y^2 = \underline{(x - y)^2} - xy - y^2$.

So next, we can get: $x(x^2 - 3xy) = x\{(x - y)^2 - xy - y^2\} = x(x - y)^2 - x^2y - xy^2$.

And next, moving on to the other part, $-3xy + y^2$, we can first, put it the way below:

$-3xy + y^2 = x^2 - x^2 - 2xy - xy + y^2 = \underline{x^2 - 2xy + y^2} - xy - x^2 = \underline{(x - y)^2} - xy - x^2$.

Thus next, we get: $3xy^2 - y^3 = -y(-3xy + y^2) = -y(x - y)^2 + xy^2 + x^2y$.

Now, putting threads together, we get:

$P = x^3 - 3x^2y + 3xy^2 - y^3 = x(x^2 - 3xy) - y(-3xy + y^2) = x(x^2 - 3xy) + \{-y(-3xy + y^2)\}$

$= \{x(x-y)^2 - x^2y - xy^2\} + \{-y(x-y)^2 + xy^2 + x^2y\}$

$= x(x-y)^2 - y(x-y)^2 - x^2y - xy^2 + xy^2 + x^2y = x(x-y)^2 - y(x-y)^2$, where $(x-y)^2$ is common to $x(x-y)^2$ and $-y(x-y)^2$, and therefore, is a divisor, that is, a factor of P.

So we get: $P = x(x-y)^2 - y(x-y)^2 = (x-y)^2(x-y) = (x-y)^3$, which is $(x-y)$ to the third power, and thus, is: $(x-y)(x-y)(x-y)$.

And we can say that P is a product of three prime factors, which are three of $(x-y)$s.

In short:

$x^3 - 3x^2y + 3xy^2 - y^3 = (x^3 - 3x^2y) + (3xy^2 - y^3) = x(x^2 - 3xy) - (-3xy + y^2)y$.

First, $x^2 - 3xy = x^2 - 2xy - xy + y^2 - y^2 = x^2 - 2xy + y^2 - xy - y^2 = (x-y)^2 - xy - y^2$.

So we get: $x(x^2 - 3xy) = x(x-y)^2 - x^2y - xy^2$.

Next, $-3xy + y^2 = x^2 - x^2 - 2xy - xy + y^2 = x^2 - 2xy + y^2 - xy - x^2 = (x-y)^2 - xy - x^2$.

So we get: $-y(-3xy + y^2) = -y(x-y)^2 + xy^2 + x^2y$.

Thus, we get:

$x(x^2 - 3xy) - y(-3xy + y^2) = x(x^2 - 3xy) + \{-y(-3xy + y^2)\}$

$= \{x(x-y)^2 - x^2y - xy^2\} + \{-y(x-y)^2 + xy^2 + x^2y\}$

$= x(x-y)^2 - y(x-y)^2 - x^2y - xy^2 + xy^2 + x^2y = x(x-y)^2 - y(x-y)^2$

$= (x-y)^2(x-y) = (x-y)^3$.

And we can have a sequence as follows:

$x - y = x - y,$

$(x - y)^2 = x^2 - 2xy + y^2,$

$(x - y)^3 = x^3 - 3x^2y + 3xy^2 - y^3,$

$(x - y)^4 = x^4 - 4x^3y + 6x^2y^2 - 4xy^3 + y^4,$

$(x - y)^5 = x^5 - 5x^4y + 10x^3y^2 - 10x^2y^3 + 5xy^4 - y^5,$

...

What then, are the expansions of $(x - y)^{1000}$ and $(x - y)^k$ where k is a nonnegative integer?

Suggestions or Solutions

To the **Problem** in the Example **1**

Assuming that $a^2 + b^2 + 2ab = K^2$, put K in terms of a and b.

$$(a + b)^2 = (a + b)(a + b) = a^2 + ab + ab + b^2 = a^2 + 2ab + b^2.$$

So $K^2 = (a + b)^2$. Thus, $K = \pm(a + b)$.

If not quite sure of the idea behind the processes above, follow the steps below:

Suppose first, **a** and **b** are lengths of two line segments.

Then, a^2 and b^2 indicate areas of two different squares, and $2ab = ab + ab$ indicates the sum of areas of two same rectangles.

And suppose now, $a > b$.

Then, we can put together the two squares and two rectangles the way below.

Fig. 1.0

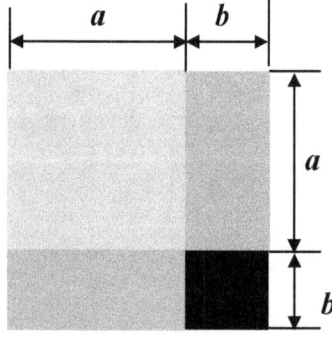

Then, we can see that the sum of all the areas of the rectangles and squares is equal to an area of one large square, which is $(a + b)$ by $(a + b)$.

In other words, we get: $(a + b)^2 = a^2 + ab + ab + b^2 = a^2 + 2ab + b^2$. So we get:

$K^2 = (a + b)^2 \Rightarrow K = a + b$ or $-(a + b)$ since we have: $\{-(a + b)\}^2 = (a + b)^2$, too.

And thus, we can put K this way, too: $K = \pm(a + b)$.

Though it sounds quite natural, we want to keep in mind that if for instance, $x^2 = y^2$, we get: $x = y$ or $-y$, and not just $x = y$ only. In short, we have: $x^2 = y^2 \Rightarrow x = \pm y$.

What if we set: $a^2 + b^2 - 2ab = K^2$, and want to put K in terms of a and b?

Let's suppose again, a and b are lengths of two line segments.

Then, a^2 and b^2 indicate areas of two different squares, and $2ab = ab + ab$ indicates the sum of areas of two same rectangles.

Suppose now again, $a > b$.

Suppose also, one rectangle a by b is put on top of the square a by a the way below:

Fig. 1.1

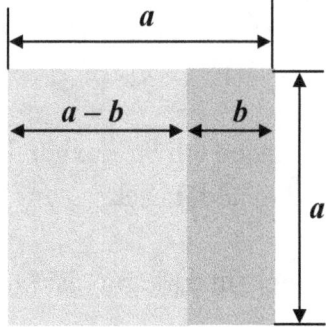

Suppose next, the other rectangle a by b is put on top of the two quadrangles as shown in the figure below on the right.

Fig. 1.2

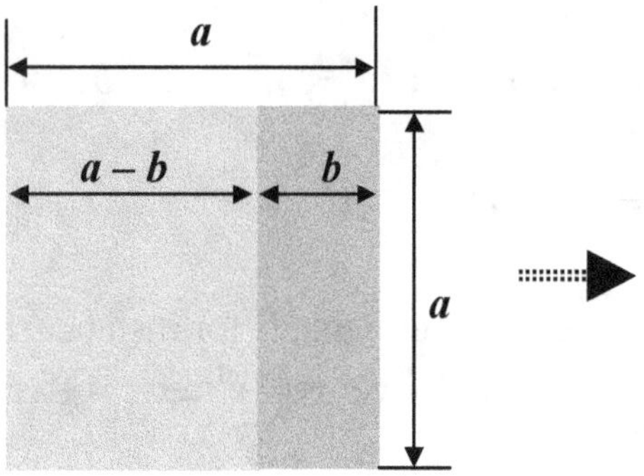

Fig. 1.3

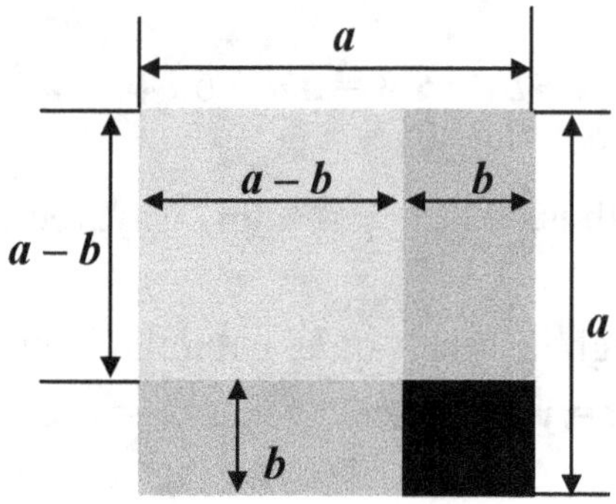

Then, we can see that the area the two same rectangles overlap is b^2.

That is, the area of the small square in black is b^2.

Also, we can see another square, which is: $(a - b)$ by $(a - b)$, so its area is: $(a - b)^2$.

And thus, we now have three different squares: small, medium, and large.

The large is: a by a, the medium is: $(a - b)$ by $(a - b)$, and the small is: b by b.

Let's now, express the medium square in terms of the two rectangles a by b, together with the other two squares. That is, we want to put $(a - b)^2$ in terms of ab, a^2, and b^2.

Then, we get: $(a - b)^2 = a^2 - ab - ab + b^2$. How come we have to add b^2, though?

Subtracting twice the rectangle a by b, from the large square a by a, we get the medium square less the small square, which is b^2. So we need to add it back.

And in the figure on the right hand side above, a rectangle in dark gray is: $(a - b)$ by b.

So the area of the medium square $(a - b)$ by $(a - b)$ is as follows:
$a^2 - (a - b)b - (a - b)b - b^2 = a^2 - ab + b^2 - ab + b^2 - b^2 = a^2 - ab - ab + b^2$.

Thus, we get: $(a - b)^2 = a^2 - ab - ab + b^2 = a^2 - 2ab + b^2 \Rightarrow (a - b)^2 = a^2 - 2ab + b^2$.

So we get: $K^2 = (a - b)^2 \Rightarrow K = a - b$ or $-(a - b)$, because $\{-(a - b)\}^2 = (a - b)^2$, too.

And thus, we can put K this way, too: $K = \pm(a - b)$.

Why do we do this example though?

• First, practicing polynomial factorizations, we can develop and improve algebra skills.

• And next, doing algebra, or setting up equations, we don't want to just put expressions and equal signs. We want to know the meanings of those expressions.

So for instance, using factorizations, we can show how an area of a square changes as we increase or decrease each side by the same amount.

And the factorizations are: $(a + b)^2 = a^2 + 2ab + b^2$, and $(a - b)^2 = a^2 - 2ab + b^2$.

So if each side of the square a by a is increased by the amount of b, the amount increased is: $2ab + b^2$.

And if each side of the square a by a is decreased by the amount of b, the amount decreased is: $2ab - b^2$, because we have: $a^2 - 2ab + b^2 = a^2 - (2ab - b^2)$.

In short:

$(a + b)^2 = (a + b)(a + b) = a^2 + ab + ab + b^2 = a^2 + 2ab + b^2$.

So $K^2 = (a + b)^2$. Thus, $K = \pm(a + b)$.

$(a - b)^2 = (a - b)(a - b) = a^2 - ab - ab + b^2 = a^2 - 2ab + b^2$.

So $K^2 = (a - b)^2$. Therefore, $K = \pm(a - b)$.

Examples 3

Factorize the polynomials below.

0. $x^2 + (a + b)x + ab$

1. $x^2 - (a + b)x + ab$

2. $acx^2 + (ad + bc)x + bd$

3. $x^3 + (a + b + c)x^2 + (ab + bc + ca)x + abc$

Suggestions or Solutions
To the **Problem** in the Example **0**

We have: $x^2 + (a + b)x + ab$.

Factorizing it, we get:

$$x^2 + (a + b)x + ab = (x^2 + ax) + (bx + ab) = x(x + a) + b(x + a) = (x + a)(x + b).$$

If not quite sure of the idea behind the processes above, follow the steps below:

Setting first, $P = x^2 + (a + b)x + ab$, we can say that the polynomial P is similar to polynomials as $x^2 + 2xy + y^2$, which is factorized to $(x + y)^2$.

And we can say that it is similar to $x^2 - 2xy + y^2$, too, which is factorized to $(x - y)^2$. How come are they similar, though?

Let's put y into each of a and b.

Then, we get: $x^2 + (a + b)x + ab = x^2 + (y + y)x + yy = x^2 + 2xy + y^2$.

And we know: $x^2 + 2xy + y^2$ is factorized to $(x + y)(x + y)$, which is: $(x + y)^2$.

Let's this time, put each of a and b into y in $(x + y)(x + y)$.
Then, we get: $(x + a)(x + b)$.

Thus, we can expect that the factorization of P is similar to that of $x^2 + 2xy + y^2$.   So?

So we may want to try doing the factorization of P the way we did to: $x^2 + 2xy + y^2$.

To begin with, we want to break the term $(a + b)x$ into two parts.

Then, we can get: $P = x^2 + (a + b)x + ab = x^2 + ax + bx + ab = (x^2 + ax) + (bx + ab)$.

What then, can we get?

Then, x is a divisor common in the first part, and b is a divisor common in the second.

So we get: $P = x(x + a) + b(x + a)$, where $x + a$ is common to $x(x + a)$ and $b(x + a)$.

Thus, we get: $P = (x + a)(x + b)$.

And we know: $(x + a)$ is prime, and so is $(x + b)$.

So P is (fully) factorized to $(x + a)(x + b)$.

In short:

$x^2 + (a + b)x + ab = (x^2 + ax) + (bx + ab) = x(x + a) + b(x + a) = (x + a)(x + b)$.

Suggestions or Solutions
To the **Problem** in the Example **1**

We have: $x^2 - (a + b)x + ab$.

Factorizing it, we get:

$$x^2 - (a + b)x + ab = (x^2 - ax) + (-bx + ab) = x(x - a) - b(x - a) = (x - a)(x - b).$$

If not quite sure of the idea behind the processes above, follow the steps below:

Setting first, $P = x^2 - (a + b)x + ab$, we can say that the polynomial P is quite similar to the one in the problem 4 above.

And in fact, it is no other than $x^2 + (a + b)x + ab$.   How come?

In $x^2 + (a + b)x + ab$, replacing x with $-x$, we get: $x^2 - (a + b)x + ab$.

And we know: $x^2 + (a + b)x + ab = (x + a)(x + b)$.

So replacing x with $-x$, we get: $(-x + a)(-x + b) = -(x - a)(-1)(x - b) = (x - a)(x - b)$.

And we can get the same the way below, too:

In $x^2 + (a + b)x + ab$, replacing a with $-a$, and b with $-b$, we get: $x^2 - (a + b)x + ab$.

And also, in $(x + a)(x + b)$, replacing a with $-a$, and b with $-b$, we get: $(x - a)(x - b)$.

And thus, recognizing patterns quickly, we can get to the solutions quickly, too.
How can we then, recognize patterns fast?

Understanding the principles, and doing practices, we can get such a caliber in pattern recognition.

Now, the polynomial P is similar to a polynomial $x^2 + (a + b)x + ab = (x + a)(x + b)$.

So let's factorize P the way we did to the polynomial $x^2 + (a + b)x + ab$.

To begin with, breaking P into two parts, we can get:

$P = x^2 - (a + b)x + ab = (x^2 - ax) + (-bx + ab)$, where we can see x is common in the first part, and $-b$ is common in the second part.

So we can get: $P = (x^2 - ax) + (-bx + ab) = x(x - a) + \{-bx + (-a)(-b)\}$

$= x(x - a) + (-b)\{x + (-a)\} = x(x - a) - b(x - a)$, where $x - a$ is common.

Thus, we get: $P = x(x - a) - b(x - a) = (x - a)(x - b)$.

In short:

$x^2 - (a + b)x + ab = (x^2 - ax) + (-bx + ab) = x(x - a) - b(x - a) = (x - a)(x - b)$.

So for instance, we can put some polynomials the way below:
$x^2 - 5x + 6 = (x^2 - 2x) + (-3x + 6) = x(x - 2) - 3(x - 2) = (x - 2)(x - 3)$.

More specifically:
$x^2 - (2 + 3)x + 2 \cdot 3 = (x^2 - 2x) + (-3x + 2 \cdot 3) = x(x - 2) - 3(x - 2) = (x - 2)(x - 3)$.

So in short: $x^2 - 5x + 6 = x^2 - (2 + 3)x + 2 \cdot 3 = (x - 2)(x - 3)$.

And we have this, too:
$x^2 + (a + b)x + ab = (x^2 + ax) + (bx + ab) = x(x + a) + b(x + a) = (x + a)(x + b)$.

So for instance, $x^2 + 5x + 6 = (x^2 + 2x) + (3x + 6) = x(x + 2) + 3(x + 2) = (x + 2)(x + 3)$.

And more specifically:

$x^2 + (2 + 3)x + 2 \cdot 3 = (x^2 + 2x) + (3x + 2 \cdot 3) = x(x + 2) + 3(x + 2) = (x + 2)(x + 3)$.

So in short: $x^2 + 5x + 6 = x^2 + (2 + 3)x + 2 \cdot 3 = (x + 2)(x + 3)$.

What then, about this: $x^2 - x - 6$?

We can put it the way below:

$x^2 - x - 6 = (x^2 + 2x) + (-3x - 6) = x(x + 2) - 3(x + 2) = (x + 2)(x - 3)$.

And more specifically:

$x^2 + (2 - 3)x + 2 \cdot (-3) = (x^2 + 2x) + \{-3x + 2 \cdot (-3)\} = x(x + 2) + (-3)(x + 2) = (x + 2)(x - 3)$.

So in short: $x^2 - x - 6 = x^2 + (2 - 3)x + 2 \cdot (-3) = (x + 2)(x - 3)$.

And we will get to the details on the processes above in one of the other sets of examples.

Suggestions or Solutions
To the **Problem** in the Example **2**

We have: $acx^2 + (ad + bc)x + bd$.

Factorizing it, we get:

$$acx^2 + (ad + bc)x + bd = (acx^2 + adx) + (bcx + bd) = ax(cx + d) + b(cx + d)$$
$$= (cx + d)(ax + b).$$

If not quite sure of the idea behind the processes above, follow the steps below:

Setting first, $P = acx^2 + (ad + bc)x + bd$, we can say that the polynomial P is just a bit different from the one in the problem 5 above, that is, P is quite similar to the one.

So after all, P is similar to $x^2 + 2xy + y^2$. How come?

First, we have: $x^2 + 2xy + y^2 = (x + y)(x + y)$, which is $(x + y)^2$.

And in $(x + y)(x + y)$, putting b and d each into each y, and also, putting ax and cx each into each x, we can get: $(ax + b)(cx + d)$.

And next, expanding the above, we get: $acx^2 + (ad + bc)x + bd$.

So the polynomial P is similar to the polynomial $x^2 + 2xy + y^2 = (x + y)^2$ in structure.

So let's factorize P the way we did to the polynomial $x^2 + 2xy + y^2$.

To begin with, breaking the polynomial P into two parts, we can get:

$P = acx^2 + (ad + bc)x + bd = (acx^2 + adx) + (bcx + bd)$, where we can see ax is common in the first part, and b is common in the second part.

So we get: $P = (acx^2 + adx) + (bcx + bd) = ax(cx + d) + b(cx + d)$, where we can see that $cx + d$ is common to $ax(cx + d)$ and $b(cx + d)$.

Thus, we get: $P = ax(cx + d) + b(cx + d) = (cx + d)(ax + b)$.

In short:

$acx^2 + (ad + bc)x + bd = (acx^2 + adx) + (bcx + bd) = ax(cx + d) + b(cx + d)$

$= (cx + d)(ax + b)$.

So for instance, we can put some polynomials the way below:

$6x^2 + 31x + 35 = (6x^2 + 10x) + (21x + 35) = 2x(3x + 5) + 7(3x + 5)$

$= (3x + 5)(2x + 7)$.

A bit more specifically:

$6x^2 + (10 + 21)x + 35 = (6x^2 + 10x) + (21x + 35) = 2x(3x + 5) + 7(3x + 5)$

$= (3x + 5)(2x + 7)$.

And more specifically:

$2 \cdot 3x^2 + (2 \cdot 5 + 7 \cdot 3)x + 7 \cdot 5 = (2 \cdot 3x^2 + 2 \cdot 5x) + (7 \cdot 3x + 7 \cdot 5) = 2x(3x + 5) + 7(3x + 5)$

$= (3x + 5)(2x + 7)$.

So in short, $6x^2 + 31x + 35 = 6x^2 + (10 + 21)x + 35 = 2 \cdot 3x^2 + (10 + 21)x + 7 \cdot 5$

$= (2 \cdot 3x^2 + 2 \cdot 5x) + (7 \cdot 3x + 7 \cdot 5) = 2x(3x + 5) + 7(3x + 5) = (3x + 5)(2x + 7)$.

What then, about this: $6x^2 + 11x - 35$?

We can put it this way:

$$6x^2 + 11x - 35 = 6x^2 + (21 - 10)x - 35 = (6x^2 - 10x) + (21x - 35)$$

$$= 2x(3x - 5) + 7(3x - 5) = (3x - 5)(2x + 7).$$

And more specifically:

$$2 \cdot 3x^2 + \{7 \cdot 3 + 2 \cdot (-5)\}x + 7 \cdot (-5) = \{2 \cdot 3x^2 + 2 \cdot (-5)x\} + \{7 \cdot 3x + 7 \cdot (-5)\}$$

$$= 2x(3x - 5) + 7(3x - 5) = (3x - 5)(2x + 7).$$

So in short, $6x^2 + 11x - 35 = 2 \cdot 3x^2 + (21 - 10)x + 7 \cdot (-5) = 2 \cdot 3x^2 + (-10 + 21)x + 7 \cdot (-5)$

$$= \{2 \cdot 3x^2 + 2 \cdot (-5)x\} + \{7 \cdot 3x + 7 \cdot (-5)\} = 2x(3x - 5) + 7(3x - 5) = (3x - 5)(2x + 7).$$

And we will get to the details on the processes above in one of the other sets of examples.

Suggestions or Solutions
To the **Problem** in the Example **3**

We have: $x^3 + (a + b + c)x^2 + (ac + bc + ca)x + abc$.

Factorizing it, we can get it the way below:

$$x^3 + (a + b + c)x^2 + (ac + bc + ca)x + abc$$

$$= x\{x^2 + (a + b + c)x + (ab + bc + ca)\} + abc$$

$$= x\{x^2 + (a + b)x + cx + ab + (bc + ca)\} + abc$$

$$= x\{x^2 + (a + b)x + ab + cx + (bc + ca)\} + abc$$

$$= x\{x^2 + (a + b)x + ab\} + cx^2 + (bc + ca)x + abc.$$

Meanwhile:

$$x\{x^2 + (a + b)x + ab\} = x(x + a)(x + b).$$
$$cx^2 + (bc + ca)x + abc = c\{x^2 + (a + b)x + ab\} = c(x + a)(x + b).$$

So $x\{x^2 + (a + b)x + ab\} + cx^2 + (bc + ca)x + abc = x(x + a)(x + b) + c(x + a)(x + b)$

$$= (x + a)(x + b)(x + c).$$

Therefore, $x^3 + (a + b + c)x^2 + (ac + bc + ca)x + abc = (x + a)(x + b)(x + c)$.

If not quite sure of the idea behind the processes above, follow the steps below:

Let's set first, $P = x^3 + (a + b + c)x^2 + (ac + bc + ca)x + abc$.

We can expect that the polynomial P is similar to $x^2 + (a + b)x + ab$, which is factorized to $(x + a)(x + b)$.

That is, we can make a logical guess that $(x + a)(x + b)(x + c)$ will do.

Expanding (that is, simplifying) it, we will see that it's the one P gets factorized to.

Now, the polynomial P is similar to $x^2 + (a + b)x + ab$, which is of degree 2.

So we may want to convert P to another polynomial containing a polynomial of degree 2. Then, we can get:

$$x^3 + (a + b + c)x^2 + (ac + bc + ca)x + abc = x\{x^2 + (a + b + c)x + (ab + bc + ca)\} + abc.$$

So we now have: $P = x\{x^2 + (a + b + c)x + (ab + bc + ca)\} + abc.$

Next, we can extract the polynomial $x^2 + (a + b)x + ab$ from P above.   How?

We can put P this way: $P = x\{x^2 + (a + b)x + cx + ab + (bc + ca)\} + abc.$

So we can get: $P = x\{x^2 + (a + b)x + ab + cx + (bc + ca)\} + abc.$

And we have: $x^2 + (a + b)x + ab = (x + a)(x + b)$. So we get:

$$P = x\{(x + a)(x + b) + cx + (bc + ca)\} + abc = x(x + a)(x + b) + cx^2 + (bc + ca)x + abc.$$

What then?

We can notice that c is a divisor common to all the terms in $cx^2 + (bc + ca)x + abc$, because $(bc + ca)x = c(b + a)x.$

So taking care of $cx^2 + (bc + ca)x + abc$ for now, we can see that:

$cx^2 + (bc + ca)x + abc = c\{x^2 + (a + b)x + ab\}$, which looks quite familiar, and is equal to $c(x + a)(x + b)$.

Thus, we get: $P = x(x + a)(x + b) + c(x + a)(x + b)$, where $(x + a)(x + b)$ is a common divisor.

Therefore, we get: $P = (x + a)(x + b)(x + c).$

In short:

$x^3 + (a + b + c)x^2 + (ac + bc + ca)x + abc$

$= x\{x^2 + (a + b + c)x + (ab + bc + ca)\} + abc$

$= x\{x^2 + (a + b)x + cx + ab + (bc + ca)\} + abc$

$= x\{x^2 + (a + b)x + ab + cx + (bc + ca)\} + abc$

$= x\{x^2 + (a + b)x + ab\} + cx^2 + (bc + ca)x + abc.$

Meanwhile:

$x\{x^2 + (a + b)x + ab\} = x(x + a)(x + b).$

$cx^2 + (bc + ca)x + abc = c\{x^2 + (a + b)x + ab\} = c(x + a)(x + b).$

So $x\{x^2 + (a + b)x + ab\} + cx^2 + (bc + ca)x + abc = x(x + a)(x + b) + c(x + a)(x + b)$

$= (x + a)(x + b)(x + c).$

Therefore, $x^3 + (a + b + c)x^2 + (ac + bc + ca)x + abc = (x + a)(x + b)(x + c).$

Examples 4

Factorize the two polynomials below.

0. $x^3 - (a + b + c)x^2 + (ab + bc + ca)x - abc$

1. $a^2 + b^2 + c^2 + 2(ab + bc + ca)$

Suggestions or Solutions
To the **Problem** in the Example **0**

We have: $x^3 - (a + b + c)x^2 + (ac + bc + ca)x - abc$.

Factorizing it, we can get it the way below:

$x^3 - (a + b + c)x^2 + (ac + bc + ca)x - abc$

$= x\{x^2 - (a + b + c)x + (ab + bc + ca)\} - abc$

$= x\{x^2 - (a + b)x - cx + ab + (bc + ca)\} - abc$

$= x\{x^2 - (a + b)x + ab - cx + (bc + ca)\} - abc$

$= x\{x^2 - (a + b)x + ab\} - cx^2 + (bc + ca)x - abc$.

Meanwhile:

$x\{x^2 - (a + b)x + ab\} = x(x - a)(x - b)$.

$-cx^2 + (bc + ca)x - abc = -c\{x^2 - (a + b)x + ab\} = -c(x - a)(x - b)$.

So $x\{x^2 - (a + b)x + ab\} - cx^2 + (bc + ca)x - abc = x(x - a)(x - b) - c(x - a)(x - b)$

$= (x - a)(x - b)(x - c)$.

Therefore, $x^3 - (a + b + c)x^2 + (ac + bc + ca)x - abc = (x - a)(x - b)(x - c)$.

If not quite sure of the idea behind the processes above, follow the steps below:

Let's set first, $P = x^3 - (a + b + c)x^2 + (ac + bc + ca)x - abc$.

The polynomial P is no different from the one in the previous problem.

In the polynomial $x^3 + (a + b + c)x^2 + (ac + bc + ca)x + abc$, simply switching a, b, and c with $-a$, $-b$, and $-c$ respectively, we get P.

Now, we know:

$x^3 + (a + b + c)x^2 + (ac + bc + ca)x + abc$ gets factorized to $(x + a)(x + b)(x + c)$.

So we can readily see that P gets factorized to $(x - a)(x - b)(x - c)$.

That is, in $x^3 + (a + b + c)x^2 + (ac + bc + ca)x + abc = (x + a)(x + b)(x + c)$, changing a, b, and c with $-a$, $-b$, and $-c$ respectively, we get:

$x^3 - (a + b + c)x^2 + (ac + bc + ca)x - abc = (x - a)(x - b)(x - c)$.

Let's see though, how P can actually be factorized in a usual manner.

This time, '$x^2 - (a + b)x + ab$' is much closer to P than '$x^2 + (a + b)x + ab$', and gets factorized to $(x - a)(x - b)$, which is the solution to the problem 5 above.

So to begin with, since P is similar to $x^2 - (a + b)x + ab$, we may want to convert P to another polynomial containing a polynomial of degree 2.   Then, we get:

$x^3 - (a + b + c)x^2 + (ac + bc + ca)x - abc = x\{x^2 - (a + b + c)x + (ab + bc + ca)\} - abc$.

So we get: $P = x\{x^2 - (a + b + c)x + (ab + bc + ca)\} - abc$, which can be put this way, too: $x\{x^2 - (a + b)x - cx + ab + (bc + ca)\} - abc$.

Then, we can get: $P = x\{x^2 - (a + b)x + ab - cx + (bc + ca)\} - abc$.

And we have: $x^2 - (a + b)x + ab = (x - a)(x - b)$.   So we get:

$P = x\{(x - a)(x - b) - cx + (bc + ca)\} - abc = x(x - a)(x - b) - cx^2 + (bc + ca)x - abc$.

Now, we can notice that $-c$ is a divisor common to every term in $-cx^2 + (bc + ca)x - abc$, because $(bc + ca)x = -c(-b - a)x$.

So we can see that:

$-cx^2 + (bc + ca)x - abc = -c\{x^2 + (-a - b)x + ab\} = -c\{x^2 - (a + b)x + ab\}$, which is equal to $-c(x - a)(x - b)$.

Thus, we get: $P = x(x - a)(x - b) - c(x - a)(x - b)$, where $(x - a)(x - b)$ is a common divisor.

Therefore, we get: $P = (x - a)(x - b)(x - c)$.

In short:

$x^3 - (a + b + c)x^2 + (ac + bc + ca)x - abc$

$= x\{x^2 - (a + b + c)x + (ab + bc + ca)\} - abc$

$= x\{x^2 - (a + b)x - cx + ab + (bc + ca)\} - abc$

$= x\{x^2 - (a + b)x + ab - cx + (bc + ca)\} - abc$

$= x\{x^2 - (a + b)x + ab\} - cx^2 + (bc + ca)x - abc.$

Meanwhile:

$x\{x^2 - (a + b)x + ab\} = x(x - a)(x - b).$

$-cx^2 + (bc + ca)x - abc = -c\{x^2 - (a + b)x + ab\} = -c(x - a)(x - b).$

So $x\{x^2 - (a + b)x + ab\} - cx^2 + (bc + ca)x - abc = x(x - a)(x - b) - c(x - a)(x - b)$

$= (x - a)(x - b)(x - c).$

Therefore, $x^3 - (a + b + c)x^2 + (ac + bc + ca)x - abc = (x - a)(x - b)(x - c).$

Suggestions or Solutions
To the **Problem** in the Example 1

We have: $a^2 + b^2 + c^2 + 2(ab + bc + ca)$.

Factorizing it, we can get it the way below:

$a^2 + b^2 + c^2 + 2(ab + bc + ca) = a^2 + 2ab + b^2 + c^2 + 2(bc + ca)$

$= (a + b)^2 + c^2 + 2c(a + b) = (a + b + c)^2$ because $x^2 + 2xy + y^2 = (x + y)^2$.

$a^2 + b^2 + c^2 + 2(ab + bc + ca) = a^2 + 2ab + b^2 + c^2 + 2(bc + ca)$

$= (a + b)^2 + c^2 + 2bc + 2ca = (a + b)^2 + c^2 + bc + bc + ca + ca$

$= (a + b)^2 + c(c + a + b) + ca + bc = (a + b)^2 + c(a + b + c) + c(a + b)$

$= (a + b)^2 + c(a + b) + c(a + b + c) = (a + b)(a + b + c) + c(a + b + c)$

$= (a + b + c)(a + b + c) = (a + b + c)^2$.

$a^2 + b^2 + c^2 + 2(ab + bc + ca) = a^2 + 2ab + b^2 + c^2 + 2(bc + ca)$

$= (a + b)^2 + c^2 + 2bc + 2ca = (a + b)^2 + c^2 + bc + bc + ca + ca$

$= (a + b)^2 + c(a + b + c) + c(a + b) = (a + b)(a + b + c) + c(a + b + c)$

$= (a + b + c)(a + b + c) = (a + b + c)^2$.

If not quite sure of the idea behind the processes above, follow the steps below:

Setting first, $P = a^2 + b^2 + c^2 + 2(ab + bc + ca)$, and examining the polynomial P, we can notice that the polynomial P has $a^2 + 2ab + b^2$, which is factorized to a complete square, $(a + b)^2$, and also, that the polynomial is symmetric.

Even changing each of a, b, and c with one of the others, we get the polynomial P again. So we can expect that the factorization is much likely to end up with a complete square.

Let's see now, how the factorization can proceed.

We may want to begin with expanding (simplifying) the last term in P so that we can form $a^2 + 2ab + b^2$ in the polynomial P.

Then, we get:

$P = a^2 + b^2 + c^2 + 2(ab + bc + ca) = a^2 + 2ab + b^2 + c^2 + 2(bc + ca)$
$= (a + b)^2 + c^2 + 2(bc + ca)$, where we can see that c is a factor of $2(bc + ca)$.

So taking out c, we get: $P = (a + b)^2 + c^2 + 2c(b + a)$.

Now, suppose that $d = a + b$.
Then, we get: $P = d^2 + 2cd + c^2$, which is factorized to $(d + c)^2$.

So we now have: $P = (d + c)^2 = (a + b + c)^2$, because $d = a + b$.

- And we can get the same result the way below, too.

Beginning with the step where $P = (a + b)^2 + c^2 + 2(bc + ca)$, and expanding $2(bc + ca)$, we get:
$P = (a + b)^2 + c^2 + 2bc + 2ca = (a + b)^2 + c^2 + bc + bc + ca + ca$.

Rearranging the terms, we can get: $P = (a + b)^2 + c^2 + ca + bc + ca + bc$.

Thus, we can say that c is a factor of $c^2 + ca + bc$.   So we can take it out.

Then, we get: $P = (a + b)^2 + c(c + a + b) + ca + bc$.

Next, we can say that c is a factor of $ca + bc$, so we can take it out.

Then, we get: $P = (a + b)^2 + c(a + b + c) + c(a + b) = (a + b)^2 + c(a + b) + c(a + b + c)$.

Next, we can say that $a + b$ is a factor of $(a + b)^2 + c(a + b)$, so we can take it out.

Then, we get: $P = (a + b)(a + b + c) + c(a + b + c)$, where $a + b + c$ is a common divisor.

So $a + b + c$ is a factor, and thus, we get: $P = (a + b + c)(a + b + c) = (a + b + c)^2$.

• Let's have a look at another way.

Beginning with the step where $P = (a + b)^2 + c^2 + ca + bc + ca + bc$, we can say that c is a factor of $ca + bc$.

So we get: $P = (a + b)^2 + c^2 + c(a + b) + ca + bc$.

Then, we can say that c is a factor of $c^2 + c(a + b)$, and is a factor of $ca + bc$, too.

So we get: $P = (a + b)^2 + c(a + b + c) + c(a + b)$.

Next, we can see that $a + b$ is a factor of $(a + b)^2 + c(a + b)$.

So we get: $P = (a + b)(a + b + c) + c(a + b + c)$, where $a + b + c$ is a common divisor.

Therefore, $a + b + c$ is a factor, and thus, we get: $P = (a + b + c)(a + b + c) = (a + b + c)^2$.

Note:

$x^2 \pm 2xy + y^2 = (a \pm b)^2$ means $x^2 + 2xy + y^2 = (a + b)^2$ or $x^2 - 2xy + y^2 = (a - b)^2$, and they are frequently used in algebraic operations.

In short:

$$a^2 + b^2 + c^2 + 2(ab + bc + ca) = a^2 + 2ab + b^2 + c^2 + 2(bc + ca)$$

$$= (a + b)^2 + c^2 + 2c(a + b) = (a + b + c)^2, \text{ because } x^2 + 2xy + y^2 = (x + y)^2.$$

$$a^2 + b^2 + c^2 + 2(ab + bc + ca) = a^2 + 2ab + b^2 + c^2 + 2(bc + ca)$$

$$= (a + b)^2 + c^2 + 2bc + 2ca = (a + b)^2 + c^2 + bc + bc + ca + ca$$

$$= (a + b)^2 + c(c + a + b) + ca + bc = (a + b)^2 + c(a + b + c) + c(a + b)$$

$$= (a + b)^2 + c(a + b) + c(a + b + c) = (a + b)(a + b + c) + c(a + b + c)$$

$$= (a + b + c)(a + b + c) = (a + b + c)^2.$$

$$a^2 + b^2 + c^2 + 2(ab + bc + ca) = a^2 + 2ab + b^2 + c^2 + 2(bc + ca)$$

$$= (a + b)^2 + c^2 + 2bc + 2ca = (a + b)^2 + c^2 + bc + bc + ca + ca$$

$$= (a + b)^2 + c(a + b + c) + c(a + b) = (a + b)(a + b + c) + c(a + b + c)$$

$$= (a + b + c)(a + b + c) = (a + b + c)^2.$$

Examples 5

Factorize the two polynomials below.

0. $a^2 + b^2 + c^2 - 2(ab - bc + ca)$

1. $a^2 - b^2$

Suggestions or Solutions
To the Problem in the Example 0

We have: $a^2 + b^2 + c^2 - 2(ab - bc + ca)$.

Factorizing it, we can get it the way below:

$a^2 + b^2 + c^2 - 2(ab - bc + ca) = a^2 - 2ab + b^2 + c^2 - 2(-bc + ca)$

$= (a - b)^2 + c^2 - 2c(a - b) = (a - b - c)^2$, because $x^2 + 2xy + y^2 = (x + y)^2$.

$a^2 + b^2 + c^2 - 2(ab - bc + ca) = a^2 - 2ab + b^2 + c^2 - 2(-bc + ca)$

$= (a - b)^2 + c^2 + 2bc - 2ca = (a - b)^2 + c^2 + bc + bc - ca - ca$

$= (a - b)^2 - c(-c - b + a) - ca + bc = (a - b)^2 - c(a - b - c) - c(a - b)$

$= (a - b)(a - b - c) - c(a - b - c) = (a - b - c)(a - b - c) = (a - b - c)^2$.

$a^2 + b^2 + c^2 - 2(ab - bc + ca) = a^2 - 2ab + b^2 + c^2 - 2(-bc + ca)$

$= (a - b)^2 + c^2 + 2bc - 2ca = (a - b)^2 + c^2 + bc + bc - ca - ca$

$= (a - b)^2 - c(a - b) + c^2 - ca + bc = (a - b)(a - b - c) - c(-c + a - b)$

$= (a - b - c)(a - b - c) = (a - b - c)^2$.

If not quite sure of the idea behind the processes above, follow the steps below:

Setting first, $P = a^2 + b^2 + c^2 - 2(ab - bc + ca)$, we can say that the polynomial P is similar to $a^2 + b^2 + c^2 + 2(ab + bc + ca)$. Such a similarity makes life much easier if we can take advantage of it.

In the polynomial $a^2 + b^2 + c^2 + 2(ab + bc + ca)$, replacing b with $-b$, and c with $-c$, we get P.

We have $a^2 + b^2 + c^2 + 2(ab + bc + ca) = (a + b + c)^2$, which is the solution to problem 9 above.

So in $(a + b + c)^2$, putting $-b$ into b, and $-c$ into c, we get $(a - b - c)^2$, to which P gets factorized.

Let's see now, how the actual factorization can proceed.

In this case, a complete square $a^2 - 2ab + b^2 = (a - b)^2$ is closer to P than $a^2 + 2ab + b^2$.

So we may want to begin with expanding the last term in P so as to extract $a^2 - 2ab + b^2$. Then, we get:

$P = a^2 + b^2 + c^2 - 2(ab - bc + ca) = a^2 - 2ab + b^2 + c^2 - 2(-bc + ca)$, where $a^2 - 2ab + b^2$

$= (a - b)^2$, and c is a factor of $2(-bc + ca)$.

So taking out c, we get: $P = (a - b)^2 + c^2 - 2c(a - b)$.

Assuming now, $d = a - b$, we can get: $P = d^2 - 2cd + c^2$, which is factorized to $(d - c)^2$.

So we now have: $P = (d - c)^2 = (a - b - c)^2$, since $d = a - b$.

- And we can get the same result the way below, too.

Beginning with the step where $P = (a - b)^2 + c^2 - 2c(a - b)$, and expanding $-2c(a - b)$, we get:
$P = (a - b)^2 + c^2 - 2ca + 2bc = (a - b)^2 + c^2 - ca - ca + bc + bc$.

Next, rearranging the terms, we can get: $P = (a - b)^2 + c^2 - ca + bc - ca + bc$.

Then, we can say that $-c$ is a factor of $c^2 - ca + bc$, and is a factor of $-ca + bc$, too.

So we get: $P = (a - b)^2 - c(-c + a - b) - c(a - b) = (a - b)^2 - c(a - b) - c(a - b - c)$.

Next, we can say that $a - b$ is a factor of $(a - b)^2 - c(a - b)$, so we can take it out.

Then, we get: $P = (a - b)(a - b - c) - c(a - b - c)$, where $a - b - c$ is a common divisor.

So $a - b - c$ is a factor, and thus, we get: $P = (a - b - c)(a - b - c) = (a - b - c)^2$.

• Let's now, have a look at another approach.

Beginning with the step $P = (a - b)^2 + c^2 - ca + bc - ca + bc$, we can say that $-c$ is a factor of $-ca + bc$.

So we get: $P = (a - b)^2 + c^2 - c(a - b) - ca + bc$.

Next, we can say that $-c$ is a factor of $c^2 - c(a - b)$, and is a factor of $-ca + bc$, too.

So $P = (a - b)^2 - c(-c + a - b) - c(a - b)$, where $a - b$ is common to $(a - b)^2$ and $-c(a - b)$.

So we get: $P = (a - b)(a - b - c) - c(a - b - c)$, where $a - b - c$ is a common divisor.

Therefore, $a - b - c$ is a factor, and thus, we get: $P = (a - b - c)(a - b - c) = (a - b - c)^2$.

In short:

$$a^2 + b^2 + c^2 - 2(ab - bc + ca) = a^2 - 2ab + b^2 + c^2 - 2(-bc + ca)$$

$$= (a - b)^2 + c^2 - 2c(a - b) = (a - b - c)^2, \text{ because } x^2 + 2xy + y^2 = (x + y)^2.$$

$$a^2 + b^2 + c^2 - 2(ab - bc + ca) = a^2 - 2ab + b^2 + c^2 - 2(-bc + ca)$$

$$= (a - b)^2 + c^2 + 2bc - 2ca = (a - b)^2 + c^2 + bc + bc - ca - ca$$

$$= (a - b)^2 - c(-c - b + a) - ca + bc = (a - b)^2 - c(a - b - c) - c(a - b)$$

$$= (a - b)(a - b - c) - c(a - b - c) = (a - b - c)(a - b - c) = (a - b - c)^2.$$

$$a^2 + b^2 + c^2 - 2(ab - bc + ca) = a^2 - 2ab + b^2 + c^2 - 2(-bc + ca)$$

$$= (a - b)^2 + c^2 + 2bc - 2ca = (a - b)^2 + c^2 + bc + bc - ca - ca$$

$$= (a - b)^2 - c(a - b) + c^2 - ca + bc = (a - b)(a - b - c) - c(-c + a - b)$$

$$= (a - b - c)(a - b - c) = (a - b - c)^2.$$

Suggestions or Solutions
To the **Problem** in the Example **1**

We have: $a^2 - b^2$.

Let's set first, $P = a^2 - b^2$.

This time, it looks quite clear that no divisor is common to all the terms in the polynomial P, and thus, P has no divisor other than 1 and itself, so is P prime?

Of course, not, so P can be factorized.

We can notice that P is similar to a polynomial $x^2 + 2xy + y^2 = (x + y)^2$.

During the factorization process of $x^2 + 2xy + y^2$, we get:

$$x^2 + 2xy + y^2 = x^2 + xy + xy + y^2 = x(x + y) + y(x + y) = \dots$$

Thus, we can expect that a term ab can be used for the factorization of P the way we use a term xy factorizing $x^2 + 2xy + y^2$.

How can we use ab, though? The polynomial P does not have ab, does it?

Of course, it doesn't. We can still use it, though.   How?

We can add it in, and then, subtract it out.

Then, we get: $P = a^2 - b^2 = a^2 - b^2 + ab - ab = a^2 + ab - b^2 - ab = (a^2 + ab) - (b^2 + ab)$.

Thus, we can see that a is a factor of $a^2 + ab$, and that b is that of $b^2 + ab$.

So we get: $P = (a^2 + ab) - (b^2 + ab) = a(a + b) - b(a + b)$, where $a + b$ is a common divisor.

Therefore, $a + b$ is a factor, and thus, we get: $P = (a + b)(a - b)$.

In short:

$$a^2 - b^2 = a^2 + ab - b^2 - ab = a(a + b) - b(a + b) = (a + b)(a - b).$$

Note:

Quite often, we take for identities all the factorizations in this set of examples.
So they are often called factorization identities.
Besides, they are frequently called factorization formulas, too.
The factorization identity $a^2 - b^2 = (a + b)(a - b)$ is probably one of the most often used when we do algebra.

And putting together all the other factorization identities covered until this example, we have:

$$x^2 + 2xy + y^2 = (x + y)^2. \qquad x^2 - 2xy + y^2 = (x - y)^2.$$

$$x^3 + 3x^2y + 3xy^2 + y^3 = (x + y)^3. \qquad x^3 - 3x^2y + 3xy^2 - y^3 = (x - y)^3.$$

$$x^2 + (a + b)x + ab = (x + a)(x + b). \qquad x^2 - (a + b)x + ab = (x - a)(x - b).$$

$$acx^2 + (ad + bc)x + bd = (ax + d)(cx + b).$$

$$x^3 + (a + b + c)x^2 + (ab + bc + ca)x + abc = (x + a)(x + b)(x + c).$$

$$x^3 - (a + b + c)x^2 + (ab + bc + ca)x - abc = (x - a)(x - b)(x - c).$$

$$a^2 + b^2 + c^2 + 2(ab + bc + ca) = (a + b + c)^2.$$

Examples 6

Factorize the two polynomials below.

0. $u^2 + 15u + 56$

1. $x^2 - 3x - 54$

Suggestions or Solutions
To the **Problem** in the Example 0

We have: $u^2 + 15u + 56$.

In this example, we have several things to discuss besides the solution.
And the things are about growing your algebra, which is the purpose of this book.
That is to say that they are for the enhancement of your algebra skill.

To begin with, we are going to get to the solution first.

Doing a factorization, we find all the prime factors applicable, and put them in a form of a product.   How then, do we find factors?

Finding factors, we find divisors, because factor are divisors.
So factorizing a polynomial, we want to find the divisors of the polynomial.
How then, can we find a divisor of a polynomial?

When finding a divisor of a polynomial, we want to find a divisor that can divide each and every term in the polynomial.
That is, finding a divisor of a polynomial, we find a divisor common to all the terms in the polynomial.   So how do we call such a divisor?

Such a divisor is called a common divisor, and does divide the polynomial.
So the common divisor is a factor of the polynomial.

And thus, finding a factor of a polynomial, we want to find a divisor common to all the terms in the polynomial.

In short, finding a factor of a polynomial, we want to find a common divisor.

It's not the case though, every polynomial shows us such a common divisor.
And the polynomial in this example, does not show it to us.
What then, do we do?

In such a case, we want to try modifying the polynomial so that we can find a common divisor.

So setting first, $P = u^2 + 15u + 56$, we can modify P the way below:

$$P = u^2 + 15u + 56 = (u^2 + 8u) + (7u + 56).$$

Then, we can take P as a polynomial made of two terms.
One is: $(u^2 + 8u)$, and the other is: $(7u + 56)$.

What then, can we notice in the each of the two terms above?

We can notice that u is common in the first term $(u^2 + 8u)$, and that 7 is common in the second term $(7u + 56)$, since $56 = 7 \cdot 8$.   So what?

So next, we can get: $P = (u^2 + 8u) + (7u + 56) = u(u + 8) + 7(u + 8)$.   What then?

Then, we can see that $(u + 8)$ is common to $u(u + 8)$ and $7(u + 8)$, and thus, is the common divisor we want.   That is, $(u + 8)$ is a factor of the polynomial P.

And thus, we get: $P = u(u + 8) + 7(u + 8) = (u + 8)(u + 7)$.

In short, $u^2 + 15u + 56 = u^2 + (8 + 7)u + 7 \cdot 8 = u^2 + 7u + 8u + 7 \cdot 8 = u(u + 7) + 8(u + 7)$

$= (u + 7)(u + 8)$.

How can we quickly see though, if the polynomial P can be modified the way above?

- Fast Factorization

Examining the polynomial $P = u^2 + 15u + 56$, we can notice that the polynomial P is similar to a polynomial as follows:

$x^2 + (a + b)x + ab$, which is factorized to $(x + a)(x + b)$.

Then in $x^2 + (a + b)x + ab$, replacing: x with u, and assuming $a + b = 15$, and $ab = 56$, we get: $u^2 + 15u + 56$, which is the polynomial P.

So finding two integers, the sum of which 15, and the product of which is 56, we get a and b. Then, we can get the factorization of P, because P is factorized to $(u + a)(u + b)$.

How then, can we find such two integers, that is, a and b?

First, either of the two integers is a, and the other is b.
And next, by trial-and-error, we can find the two.
We don't just try random integers though.   What then, do we try?

We know: $ab = 56$. So a and b are divisors of 56.

So we can set: $ab = 1 \cdot 56 = 2 \cdot 28 = 4 \cdot 14 = 8 \cdot 7$.
And also, we can set: $ab = (-1) \cdot (-56) = (-2) \cdot (-28) = (-4) \cdot (-14) = (-8) \cdot (-7)$.

And thus, (a, b) is one of the eight cases as follows:

$(1, 56)$, $(2, 28)$, $(4, 14)$, $(8, 7)$, $(-1, -56)$, $(-2, -28)$, $(-4, -14)$, and $(-8, -7)$.

And of the eight cases above, the fourth one, that is, $(8, 7)$ satisfies: $a + b = 15$.
So we get: $P = u^2 + 15u + 56 = (u + 8)(u + 7)$.
And we know: $u + 7$ is prime, and so is $u + 8$.
So we can now say that $u^2 + 15u + 56$ is fully factorized to $(u + 7)(u + 8)$.

And thus, together with the fact that a and b are divisors of 56, by trial-and-error, we can find a and b so that the sum of a and b is 15.

What if however, no case satisfies: $a + b = 15$?

Then, the polynomial is not factorable within the scope of integers.
If a polynomial is factorable within the scope of integers, we can get its factors where all the coefficients are integers.

For instance, if a polynomial has as factors $2x + 1$, $x - 1$, and $3x - 2$, the polynomial is factorable within the scope of integers, since integers are used as all the numeric terms and all the coefficients in all the factors.

If however, a factor of a polynomial is: $\sqrt{2}x - 2$ or $x + \sqrt{2}$, the polynomial is not factorable within the scope of integers, since integers cannot be use as all the coefficients and all the numeric terms.

What then, about a polynomial that has as factors $x + \frac{2}{3}$, $\frac{3}{2}x + 1$, or $\frac{2}{3}x + \frac{1}{2}$?

We can put the factors the way below:
$x + \frac{2}{3} = \frac{1}{3}(3x + 2)$, $\frac{3}{2}x + 1 = \frac{1}{2}(3x + 2)$, and $\frac{2}{3}x + \frac{1}{2} = \frac{1}{6}(4x + 3)$.

Then, all the coefficients and the numeric terms are all integers.
And in high school math, covered are factorizations within the scope of integers.

Now, the purpose of doing polynomial factorizations is to develop and enhance your algebra skill. So you may want to do some more algebra with this factorization.

Finding in fact, a and b algebraically, we want to solve a system of the two equations where $a + b = 15$, and $ab = 56$.

The system is though, no other than factorizing the polynomial P.   How come?

Solving the system of equations above, we will end up with a quadratic equation, which is no other than setting the polynomial $P = 0$.   How come?

First, we have: $a + b = 15$, so we get: $b = 15 - a$.   Thus, we get: $ab = a(15 - a) = 56$.

So we get: $a(15 - a) = 15a - a^2 = 56 \Rightarrow a^2 - 15a + 56 = 0$.

And we can factorize $a^2 - 15a + 56$ to $(a - 7)(a - 8)$.

That is, we get: $a^2 - 15a + 56 = (a - 7)(a - 8)$.

So we get: $a^2 - 15a + 56 = 0 \Rightarrow (a - 7)(a - 8) = 0 \Rightarrow a = 7$ or 8.

And we have: $b = 15 - a$.   So we get:

$a = 7 \Rightarrow b = 15 - a = 15 - 7 = 8$.

$a = 8 \Rightarrow b = 15 - a = 15 - 8 = 7$.

And thus, the solution to the system is: $a = 7$ and $b = 8$, or $a = 8$ and $b = 7$.

Next, we get: $P = 0 \Rightarrow u^2 + 15u + 56 = 0$.

And we can factorize $u^2 + 15u + 56$ to $(u + 7)(u + 8)$.

That is, we get: $u^2 + 15u + 56 = (u + 7)(u + 8)$.

So we get: $P = 0 \Rightarrow u^2 + 15u + 56 = 0 \Rightarrow (u + 7)(u + 8) = 0 \Rightarrow u = -7$ or -8.

That is, $P = 0 \Rightarrow u = -7$ or -8.

And thus, solving the system of equations where $a + b = 15$, and $ab = 56$, we just get back to the very problem in this example, that is, factorizing the polynomial P. That's because the values of $-a$ and $-b$ are the solution to $P = 0$.   How come?

The equation $P = 0$ is: $u^2 + 15u + 56 = 0$, and can be put this way: $u^2 + (a + b)u + ab = 0$.

And we have: $u^2 + (a + b)u + ab = (u + a)(u + b)$.

So the equation $u^2 + 15u + 56 = 0$ is now the same as $(u + a)(u + b) = 0$.

And solving $(u + a)(u + b) = 0$, we get: $u = -a$ or $-b$ because $u + a = 0$ or $u + b = 0$.

Thus, the values of $-a$ and $-b$ are the solution to the equation $P = 0$.

Now, we have two typical approaches to a factorization.

One is a logical guessing, which has been covered above, and the other is a straight factorization, which is in this case, using this identity: $(s + t)^2 = s^2 + 2st + t^2$.
How then, can we use the identity above?

Referring to the identity, we can put first, $P = u^2 + 15u + 56$ into the form of $s^2 + 2st + t^2$.

And next, we can convert P into a polynomial with a binomial squared as $(u + v)^2$, where v is a number. And a binomial is a polynomial that is made of two terms as $x + 1$, $2x + c$, or $x + 2y$.

And the conversion can proceed the way as follows.

To begin with, we can change P the way below:

$(1)\ P = u^2 + 15u + 56 = u^2 + 2 \cdot \frac{15}{2} \cdot u + (\frac{15}{2})^2 - (\frac{15}{2})^2 + 56$.

Next, we can change the first three terms the way below:

$u^2 + 2 \cdot \frac{15}{2} \cdot u + (\frac{15}{2})^2 = (u + \frac{15}{2})^2$.

And next, in the expression (1) above, we can change the last two terms the way below:

$56 - (\frac{15}{2})^2 = 56 - \frac{1}{4}(150 + 50 + 25) = 56 - \frac{225}{4} = \frac{56 \cdot 4 - 225}{4} = \frac{1}{4}(200 + 24 - 225) = -\frac{1}{4}$.

So we get: $P = u^2 + 15u + 56 = (u + \frac{15}{2})^2 - \frac{1}{4}$, which has a binomial squared.   So what?

We have this identity, too: $s^2 - t^2 = (s - t)(s + t)$.

So next, using a fact that $\frac{1}{4} = (\pm\frac{1}{2})^2$, together with the identity above, we get:

$P = (u + \frac{15}{2})^2 - \frac{1}{4} = (u + \frac{15}{2})^2 - (\pm\frac{1}{2})^2 = \{(u + \frac{15}{2}) + (\pm\frac{1}{2})\}\{(u + \frac{15}{2}) - (\pm\frac{1}{2})\}$.

We can however, just set: $P = \{(u + \frac{15}{2}) + \frac{1}{2}\}\{(u + \frac{15}{2}) - \frac{1}{2}\}$.   How come?

$$P = \{(u + \tfrac{15}{2}) + (\pm\tfrac{1}{2})\}\{(u + \tfrac{15}{2}) - (\pm\tfrac{1}{2})\} = \{(u + \tfrac{15}{2}) \pm \tfrac{1}{2}\}\{(u + \tfrac{15}{2}) \mp \tfrac{1}{2}\}$$

$$= \{(u + \tfrac{15}{2}) + \tfrac{1}{2}\}\{(u + \tfrac{15}{2}) - \tfrac{1}{2}\} \text{ or } \{(u + \tfrac{15}{2}) - \tfrac{1}{2}\}\{(u + \tfrac{15}{2}) + \tfrac{1}{2}\}, \text{ both of which are the same.}$$

So next, we get: $P = \{(u + \tfrac{15}{2}) + \tfrac{1}{2}\}\{(u + \tfrac{15}{2}) - \tfrac{1}{2}\} = (u + \tfrac{16}{2})(u + \tfrac{14}{2}) = (u + 8)(u + 7)$.

Thus, we get: $P = u^2 + 15u + 56 = (u + 7)(u + 8)$.

In short, we get:

$$P = u^2 + 15u + 56 = u^2 + 2 \cdot \tfrac{15}{2} \cdot u + (\tfrac{15}{2})^2 - (\tfrac{15}{2})^2 + 56 = (u + \tfrac{15}{2})^2 - \tfrac{1}{4}$$

$$\{(u + \tfrac{15}{2}) + \tfrac{1}{2}\}\{(u + \tfrac{15}{2}) - \tfrac{1}{2}\} = (u + \tfrac{16}{2})(u + \tfrac{14}{2}) = (u + 8)(u + 7).$$

- So what else can we get doing the factorization the way above?

We can solve for x the quadratic equation $ax^2 + bx + c = 0$ the way below.

First, we get: $ax^2 + bx + c = a(x^2 + \tfrac{b}{a}x) + c = a\{x^2 + 2 \cdot \tfrac{b}{2a} \cdot x + (\tfrac{b}{2a})^2 - (\tfrac{b}{2a})^2\} + c$

$$= a\{x^2 + 2 \cdot \tfrac{b}{2a} \cdot x + (\tfrac{b}{2a})^2\} - a(\tfrac{b}{2a})^2 + c = a(x + \tfrac{b}{2a})^2 - \tfrac{b^2}{4a} + c = 0.$$

Next, taking advantage of the identity $p^2 - q^2 = (p - q)(p + q)$, we get:

$$a(x + \tfrac{b}{a})^2 - \tfrac{b^2}{4a} + c = 0 \Rightarrow (x + \tfrac{b}{a})^2 - \tfrac{b^2}{4a^2} + \tfrac{c}{a} = (x + \tfrac{b}{2a})^2 - \tfrac{b^2 - 4ac}{4a^2} = 0$$

$$\Rightarrow \{(x + \tfrac{b}{2a}) + \tfrac{\sqrt{b^2 - 4ac}}{2a}\}\{(x + \tfrac{b}{2a}) - \tfrac{\sqrt{b^2 - 4ac}}{2a}\} = (x + \tfrac{b}{2a} + \tfrac{\sqrt{b^2 - 4ac}}{2a})(x + \tfrac{b}{2a} - \tfrac{\sqrt{b^2 - 4ac}}{2a})$$

$$\Rightarrow \{(x + \tfrac{b + \sqrt{b^2 - 4ac}}{2a})(x + \tfrac{b - \sqrt{b^2 - 4ac}}{2a}) = 0.$$

In other words, we get: $(x + \tfrac{b}{2a})^2 - \tfrac{b^2 - 4ac}{4a^2} = (x + \tfrac{b + \sqrt{b^2 - 4ac}}{2a})(x + \tfrac{b - \sqrt{b^2 - 4ac}}{2a}) = 0.$

Therefore, the solution to the equation $ax^2 + bx + c = 0$ is as follows:

$$x = -\frac{b+\sqrt{b^2-4ac}}{2a} \ \text{ or } \ -\frac{b-\sqrt{b^2-4ac}}{2a}, \text{ that is, } x = -\frac{b\pm\sqrt{b^2-4ac}}{2a} = \frac{-b\pm\sqrt{b^2-4ac}}{2a}$$

$$\Rightarrow x = \frac{-b\pm\sqrt{b^2-4ac}}{2a}, \text{ which is often called the } \textit{quadratic formula.}$$

So we just now have checked how to get the quadratic formula.

And the quadratic equation $ax^2 + bx + c = 0$ is one of the major equations we have to solve or work with in high school math and college math, too.

• So let's next, take a look at some nature of quadratic equations.

Knowing the nature and going through the processes below, we can get stronger the foundation of algebra.

Now, the equation $ax^2 + bx + c = 0$ is no other than $(x+\frac{b}{2a})^2 - \frac{b^2-4ac}{4a^2} = 0$.

However, we want to note that $ax^2 + bx + c \neq (x+\frac{b+\sqrt{b^2-4ac}}{2a})(x+\frac{b-\sqrt{b^2-4ac}}{2a})$, and that

$$ax^2 + bx + c = a(x+\frac{b+\sqrt{b^2-4ac}}{2a})(x+\frac{b-\sqrt{b^2-4ac}}{2a}). \quad \text{How come?}$$

That's because dividing by a both sides of $ax^2 + bx + c = 0$, we get:

$$x^2 + \frac{b}{a}x + \frac{c}{a} = 0 \Rightarrow (x+\frac{b}{a})^2 - \frac{b^2}{4a^2} + \frac{c}{a} = (x+\frac{b}{2a})^2 - \frac{b^2-4ac}{4a^2} = (x+\frac{b+\sqrt{b^2-4ac}}{2a})(x+\frac{b-\sqrt{b^2-4ac}}{2a}) = 0.$$

So multiplying $(x+\frac{b+\sqrt{b^2-4ac}}{2a})(x+\frac{b-\sqrt{b^2-4ac}}{2a})$ by a, we get: $ax^2 + bx + c = 0$.

Thus, $ax^2 + bx + c$ gets factorized to $a(x+\frac{b+\sqrt{b^2-4ac}}{2a})(x+\frac{b-\sqrt{b^2-4ac}}{2a})$.

Meanwhile, we have:

$$a\left(x+\frac{b+\sqrt{b^2-4ac}}{2a}\right)\left(x+\frac{b-\sqrt{b^2-4ac}}{2a}\right) = \left(ax+\frac{b+\sqrt{b^2-4ac}}{2}\right)\left(x+\frac{b-\sqrt{b^2-4ac}}{2a}\right).$$

$$a\left(x+\frac{b+\sqrt{b^2-4ac}}{2a}\right)\left(x+\frac{b-\sqrt{b^2-4ac}}{2a}\right) = \left(x+\frac{b+\sqrt{b^2-4ac}}{2a}\right)\left(ax+\frac{b-\sqrt{b^2-4ac}}{2}\right).$$

So $ax^2 + bx + c$ can be said to get factorized to either of the two below:

$$\left(ax+\frac{b+\sqrt{b^2-4ac}}{2}\right)\left(x+\frac{b-\sqrt{b^2-4ac}}{2a}\right) \text{ and } \left(x+\frac{b+\sqrt{b^2-4ac}}{2a}\right)\left(ax+\frac{b-\sqrt{b^2-4ac}}{2}\right).$$

In short, $ax^2 + bx + c$ gets factorized to $a\left(x+\frac{b+\sqrt{b^2-4ac}}{2a}\right)\left(x+\frac{b-\sqrt{b^2-4ac}}{2a}\right)$.

So factorizing a polynomial $ax^2 + bx + c$, we can use the quadratic formula.

- Let's for instance, factorize $3x^2 + 25x + 28$ using the formula.

We have: $a = 3$, $b = 25$, and $c = 28$. So setting $3x^2 + 25x + 28 = 0$, we get:

$$x = \frac{-b\pm\sqrt{b^2-4ac}}{2a} = \frac{-25\pm\sqrt{25^2-4\cdot3\cdot28}}{2\cdot3} = \frac{-25\pm\sqrt{625-336}}{6} = \frac{-25\pm\sqrt{289}}{6} = \frac{-25\pm17}{6} \Rightarrow x = -\frac{4}{3} \text{ or } -7.$$

Thus, the solution to $3x^2 + 25x + 28 = 0$ is: $x = -\frac{4}{3}$ or -7.

So we get: $3x^2 + 25x + 28 = 0 \Rightarrow \left(x+\frac{4}{3}\right)(x+7) = 0$.

However, the polynomial $3x^2 + 25x + 28$ does get factorized to $3\left(x+\frac{4}{3}\right)(x+7)$, and not to $\left(x+\frac{4}{3}\right)(x+7)$.

That's because dividing $3x^2 + 25x + 28 = 0$ by 3, we get: $x^2 + \frac{25}{3}x + \frac{28}{3} = 0$, and then, we get: $x^2 + \frac{25}{3}x + \frac{28}{3} = \left(x+\frac{4}{3}\right)(x+7) = 0$.

Thus, we get: $3x^2 + 25x + 28 = 3\left(x+\frac{4}{3}\right)(x+7) = (3x+4)(x+7)$.

Suggestions or Solutions
To the **Problem 1** in the Example **0**

We have: $x^2 - 3x - 54$.

Factorizing it, we can get it the way below:

$$x^2 - 3x - 54 = x^2 + 6x - 9x - 54 = x(x + 6) - 9(x + 6) = (x + 6)(x - 9).$$

If not quite sure of the idea behind the processes above, follow the steps below:

Setting first: $P = x^2 - 3x - 54$, we can say that the polynomial P is quadratic and the coefficients and numeric term are integers.

Also, P is similar to $x^2 - (a + b)x + ab$, which is factorized to $(x - a)(x - b)$.

And also, however, P can be said to be similar to $x^2 + (a + b)x + ab$, too, which gets factorized to $(x + a)(x + b)$.

That is because changing a and b with $-a$ and $-b$ respectively in $(x - a)(x - b)$, we get:

$$(x - a)(x - b) = \{x - (-a)\}\{x - (-b)\} = (x + a)(x + b).$$

So $x^2 - (a + b)x + ab$ is just about the same as $x^2 + (a + b)x + ab$.

And thus, in $x^2 + (a + b)x + ab$, assuming: $a + b = -3$, and $ab = -54$, we get: $x^2 - 3x - 54$, which is the polynomial P.

So finding two integers, the sum of which -3, and the product of which is -54, we get a and b. Then, we can get the factorization of P, because P gets factorized to $(x + a)(x + b)$, where a and b are integers.

How then, can we find such two integers, that is, a and b?

First, either of the two integers is a, and the other is b.

And next, by trial-and-error, we can find the two.

And we know: $ab = -54$.   So a and b are divisors of -54.

And by trial and error, we can try finding two integers that can get multiplied to be -54, and also, add up to -3.

If such a pair gets found, either of the two integers is a and the other is b, so P can get factorized to $(x + a)(x + b)$.

That is, assuming that a and b are integers, we want to solve a system where $ab = -54$ and $a + b = -3$.

If any, we can find two integers adding up to -3 and getting multiplied to be -54 using divisors of -54 via trial and error. Then, the two integers found are a and b.

If however, no integers satisfy the system above, that is, a and b cannot be integers, we want to consult the quadratic formula.

By trial and error with divisors of -54, we can get:
$6(-9) = -54$, and $6 + (-9) = -3$.

So either of a and b is 6, and the other is -9, and thus, P is factorized to $(x + 6)(x - 9)$.

The actual factorization process can be as follows:

$x^2 - 3x - 54 = x^2 + (6 - 9)x + 6 \cdot (-9) = x^2 + 6x - 9x + 6 \cdot (-9) = x(x + 6) - 9(x + 6)$

$= (x + 6)(x - 9)$.

In short:

$x^2 - 3x - 54 = x^2 + 6x - 9x - 54 = x(x + 6) - 9(x + 6) = (x + 6)(x - 9)$.

Examples 7

Factorize the polynomials below.

0. $y^2 + 14y + 49$

1. $t^2 + 7t + \frac{49}{4}$

2. $9t^2 + 48t + 64$

3. $\frac{4}{9}t^2 + \frac{20}{3}t + 25$

4. $49r^2 + \frac{112}{3}r + \frac{64}{9}$

Suggestions or Solutions
To the **Problem** in the Example **0**

We have: $y^2 + 14y + 49$.

Factorizing it, we get: $y^2 + 14y + 49 = y + 7y + 7y + 49 = y(y + 7) + 7(y + 7) = (y + 7)^2$.

If not quite sure of the idea behind the processes above, follow the steps below:

Frequently, a quaratic polynomial can be converted to a complete square. So for instance, we can put $u^2x^2 + 2uvx + v^2$ in a complete square this way: $u^2x^2 + 2uvx + v^2 = (ux + v)^2$.

And the simplest complete square is: $a^2 + 2ab + b^2 = (a + b)^2$.

Now, setting first, $P = y^2 + 14y + 49$, we can say that the polynomial P is quadratic, and is quite close to the simplest complete square.   So?

So putting P in the same structure as the simplest one has, we can see if it can be put in a complete square.

$P = y^2 + 14y + 49 = y^2 + 2 \cdot 7 \cdot y + 7^2 = (y + 7)^2$.

Thus, the polynomia P is factorized to $(y + 7)^2$.

And the actual factorization processes can be:

$y^2 + 14y + 49 = y^2 + 7y + 7y + 49 = y(y + 7) + 7(y + 7) = (y + 7)(y + 7) = (y + 7)^2$.

What if however, a polynomial cannot be put in a complete square?

Even if a quadratic polynomial cannot be put in a complete square, we can still use the idea of the complete square. In such a case, we reconstruct the polynomial so that it can be expressed in terms of a complete square and a nonzero numeric term.

For instance, $y^2 + 14y + 50$ can be put in $(y + 7)^2 + 1$, which is made of a complete square and a nonzero numeric term.

By the way, any real number squard (1 multiplied by the same number twice) cannot be negative. In other words, a real number squared is greater than or equal to 0.

Therefore, we get: $(y + 7)^2 \geq 0 \Rightarrow y^2 + 14y + 50 = (y + 7)^2 + 1 \geq 1$.

So we get: $y^2 + 14y + 50 \geq 1$.

In short:

$y^2 + 14y + 49 = y + 7y + 7y + 49 = y(y + 7) + 7(y + 7) = (y + 7)^2$.

Suggestions or Solutions
To the **Problem** in the Example **1**

We have: $t^2 + 7t + \frac{49}{4}$.

Factorizing it, we get can get it the way below:

$$t^2 + 7t + \frac{49}{4} = t^2 + \frac{7}{2} \cdot t + \frac{7}{2} \cdot t + (\frac{7}{2})^2 = t(t + \frac{7}{2})(\frac{7}{2})(t + \frac{7}{2}) = (t + \frac{7}{2})(t + \frac{7}{2}) = (t + \frac{7}{2})^2.$$

If not quite sure of the idea behind the processes above, follow the steps below:

Setting first, $P = t^2 + 7t + \frac{49}{4}$, we can say that P is quadratic, but could expect that P is not factorable within the scople of integers, becuase the numeric term is not an intger.

It can be the case though, P can be factorable within the scople of integers.

Many quaratic polynomials can be put in complete squares, and we can factorize readily such a polynomial. And we can quickly check to see if a quadratic polynomial can be put in a complete square evaluating the discriminant of such a polynomial.

Suppose for instance, $Q = ax^2 + bx + c$, and D is the discriminant.

Then, $D = b^2 - 4ac$, which is in fact, a part of the quadratic formula as follows:

$x = \frac{-b \pm \sqrt{b^2 - 4ac}}{2a}$. So the formula can be put this way, too: $x = \frac{-b \pm \sqrt{D}}{2a}$ where $D = b^2 - 4ac$.

Thus, we have:

$$ax^2 + bx + c = a(x - \frac{-b + \sqrt{b^2 - 4ac}}{2a})(x - \frac{-b - \sqrt{b^2 - 4ac}}{2a}) = a(x - \frac{-b + \sqrt{D}}{2a})(x - \frac{-b - \sqrt{D}}{2a}).$$

(If not sure of how to get it, refer to the example 0 in **Examples 6**.)

Therefore, if the discriminant $D = 0$, we get:

$$ax^2 + bx + c = a(x - \frac{-b}{2a})(x - \frac{-b}{2a}) = a(x + \frac{b}{2a})(x + \frac{b}{2a}) = a(x + \frac{b}{2a})^2.$$

So Q can be put in a complete square if the discriminant D is 0.

That is, a quadratic polnomial can be put in a complete square if its discriminant is 0.

What then, about $\frac{b}{2a}$, which is not an integer?

We can put it this way, too: $\frac{b}{2a} = \frac{1}{2a} \cdot b$.   So we can put $a(x + \frac{b}{2a})^2$ the way below, too:

$$a(x + \tfrac{b}{2a})^2 = a(\tfrac{2a}{2a}x + \tfrac{1}{2a}b)^2 = a\{\tfrac{1}{2a}(2ax + b)\}^2 = a(\tfrac{1}{2a})^2(2ax + b)^2 = \tfrac{1}{4a}(2ax + b)^2.$$

Now, the discriminant of P is: $7^2 - 4 \cdot 1 \cdot \frac{49}{4} = 0$. So P can be put in a complete square.

And we can get: $ax^2 + bx + c = a(x + \frac{b}{2a})^2$ if the discriminant D is 0.

Since $P = t^2 + 7t + \frac{49}{4}$, we can take 1 as a, can take 7 as b, and can take $\frac{49}{4}$ as c.

So we can get: $t^2 + 7t + \frac{49}{4} = (t + \frac{7}{2})^2$.

In fact, $t^2 + 7t + \frac{49}{4} = t^2 + 2 \cdot \frac{7}{2} \cdot t + (\frac{7}{2})^2 = (t + \frac{7}{2})^2$.

The factorization process can be as follows :

$$t^2 + 7t + \tfrac{49}{4} = t^2 + \tfrac{7}{2} \cdot t + \tfrac{7}{2} \cdot t + (\tfrac{7}{2})^2 = t(t + \tfrac{7}{2})(\tfrac{7}{2})(t + \tfrac{7}{2}) = (t + \tfrac{7}{2})(t + \tfrac{7}{2}) = (t + \tfrac{7}{2})^2.$$

And if necessary, we can put it the way below, too:

$$(t + \tfrac{7}{2})^2 = (\tfrac{2}{2}t + \tfrac{1}{2} \cdot 7)^2 = \{\tfrac{1}{2}(2t + 7)\}^2 = \tfrac{1}{4}(2t + 7)^2.$$

In short:

$$t^2 + 7t + \tfrac{49}{4} = t^2 + \tfrac{7}{2} \cdot t + \tfrac{7}{2} \cdot t + (\tfrac{7}{2})^2 = t(t + \tfrac{7}{2})(\tfrac{7}{2})(t + \tfrac{7}{2}) = (t + \tfrac{7}{2})(t + \tfrac{7}{2}) = (t + \tfrac{7}{2})^2.$$

Suggestions or Solutions
To the **Problem** in the Example **2**

We have: $9t^2 + 48t + 64$.

If a quaratic polynomial can be put in a complete square, we can readily factorize it. The discriminant can tell us immediately if it is the case.

Assumig for instance, $Q = ax^2 + bx + c$, and D is the discriminant, we get: $D = b^2 - 4ac$. And if $D = 0$, we can say that Q can be put in a complete square.

Now, setting first, $P = 9t^2 + 48t + 64$, we can say that the polynomial P is quadratic, so checking the discriminant of P, we can see if P can be put in a complete square.

So finding D for P, we get: $48^2 - 4\cdot9\cdot64 = 48^2 - (2\cdot3\cdot8)^2 = 48^2 - 48^2 = 0$.

Thus, $D = 0$, and P can be put in a complete square.

And we can put P in a complete square the way below, too:

$P = 9t^2 + 48t + 64 = (3t)^2 + 2\cdot8\cdot3t + 8^2 = (3t + 8)^2$, to which P gets factorized, of course.

And also, the factorization process can be as follows:

$9t^2 + 48t + 64 = 9t^2 + 24t + 24t + 64 = 3t(3t + 8) + 24t + 64 = 3t(3t + 8) + 8(3t + 8)$

$= (3t + 8)(3t + 8) = (3t + 8)^2$.

In short:

$9t^2 + 48t + 64 = 9t^2 + 24t + 24t + 64 = 3t(3t + 8) + 24t + 64 = 3t(3t + 8) + 8(3t + 8)$

$= (3t + 8)(3t + 8) = (3t + 8)^2$. Another way: $9t^2 + 48t + 64 = (3t)^2 + 2\cdot3\cdot8t + 8^2 = (3t + 8)^2$.

Suggestions or Solutions

To the **Problem** in the Example 3

Suppose $P = \frac{4}{9}t^2 + \frac{20}{3}t + 25$.

Factorizing it, we can get it the way below:

$$\frac{4}{9}t^2 + \frac{20}{3}t + 25 = (\frac{2}{3}t)^2 + \frac{10}{3}t + \frac{10}{3}t + 5^2 = \frac{2}{3}t(\frac{2}{3}t+5) + 5(\frac{2}{3}t+5) = (\frac{2}{3}t+5)^2.$$

If not quite sure of the idea behind the processes above, follow the steps below:

Quite often, a quaratic polynomial can be put in a complete square, and we can readily factorize such a polynomial. How then, can we see if it can be put in a complete square?

If for instance, **D** is the discriminant of $Q = ax^2 + bx + c$, then $D = b^2 - 4ac$, which is in fact, a part of the quadratic formula as follows:

$x = \frac{-b \pm \sqrt{b^2 - 4ac}}{2a}$. So the formula can be put this way, too: $x = \frac{-b \pm \sqrt{D}}{2a}$ where $D = b^2 - 4ac$.

Thus, we have:

$$ax^2 + bx + c = a(x - \frac{-b + \sqrt{b^2-4ac}}{2a})(x - \frac{-b - \sqrt{b^2-4ac}}{2a}) = a(x - \frac{-b+\sqrt{D}}{2a})(x - \frac{-b-\sqrt{D}}{2a}).$$

(If not sure of how to get it, refer to the example 0 in **Examples 6**.)

Therefore, if the discriminant $D = 0$, we get:

$$ax^2 + bx + c = a(x - \frac{-b}{2a})(x - \frac{-b}{2a}) = a(x + \frac{b}{2a})(x + \frac{b}{2a}) = a(x + \frac{b}{2a})^2.$$

That is, a quadratic polnomial can be put in a complete square if its discriminant is 0.

And we can put the polynomial Q above the way below, too:

$$ax^2 + bx + c = a(x + \tfrac{b}{2a})^2 = (\sqrt{a} \cdot x + \tfrac{\sqrt{a \cdot b}}{2a})^2, \text{ where } \sqrt{a}, \text{ and } \tfrac{\sqrt{a \cdot b}}{2a} \text{ are constant.}$$

Thus, setting $m = \sqrt{a}$, and $n = \tfrac{\sqrt{a \cdot b}}{2a}$, we can set Q this way, too: $ax^2 + bx + c = (mx + n)^2$.

So evaluating the discriminant D of P, we can quickly check to see if P can be put in a complete square.

Now, P's D is: $(\tfrac{20}{3})^2 - 4 \cdot (\tfrac{4}{9}) \cdot 25 = (\tfrac{20}{3})^2 - \tfrac{4^2 \cdot 5^2}{3^2} = (\tfrac{20}{3})^2 - (\tfrac{20}{3})^2 = 0$.

So P can be put in a complete square, and we get:

$$\tfrac{4}{9}t^2 + \tfrac{20}{3}t + 25 = (\tfrac{2}{3}t)^2 + 2 \cdot \tfrac{2}{3}t \cdot 5 + 5^2 = (\tfrac{2}{3}t + 5)^2.$$

And the factorization process can be as follows:

$$\tfrac{4}{9}t^2 + \tfrac{20}{3}t + 25 = (\tfrac{2}{3}t)^2 + \tfrac{10}{3}t + \tfrac{10}{3}t + 5^2 = \tfrac{2}{3}t(\tfrac{2}{3}t + 5) + 5(\tfrac{2}{3}t + 5) = (\tfrac{2}{3}t + 5)^2.$$

And also, if necessary, we can put it the way below, too:

$$(\tfrac{2}{3}t + 5)^2 = (\tfrac{2}{3}t + \tfrac{3 \cdot 5}{3})^2 = \{\tfrac{1}{3}(2t + 15)\}^2 = \tfrac{1}{9}(2t + 15)^2.$$

In short:

$$\tfrac{4}{9}t^2 + \tfrac{20}{3}t + 25 = (\tfrac{2}{3}t)^2 + \tfrac{10}{3}t + \tfrac{10}{3}t + 5^2 = \tfrac{2}{3}t(\tfrac{2}{3}t + 5) + 5(\tfrac{2}{3}t + 5) = (\tfrac{2}{3}t + 5)^2.$$

Suggestions or Solutions
To the **Problem** in the Example **4**

We have: $49r^2 + \frac{112}{3}r + \frac{64}{9}$.

Factorizing it, we can get it the way below :

$$P = 49r^2 + \frac{112}{3}r + \frac{64}{9} = 49r^2 + \frac{56}{3}r + \frac{56}{3}r + \frac{64}{9} = 7r(7r + \frac{8}{3}) + \frac{8}{3}(7r + \frac{8}{3}) = (7r + \frac{8}{3})^2.$$

If not quite sure of the idea behind the processes above, follow the steps below:

Setting first, $P = 49r^2 + \frac{112}{3}r + \frac{64}{9}$, we could expect that P is not factorable within the scope of integers, because in P, all the coefficients and the numeric term are not integers.

It is often the case however, we can put a quadratic polynomial into a complete square.

For instance, $(x + \frac{1}{2})^2$ is a quadratic polynomial, and is in a complete square, and is in fact, equal to $x^2 + x + \frac{1}{4}$.

And we can see if P can be put in a complete square if we get the discriminant of P, and see if its discriminant is 0.

Suppose for instance, D is the discriminat of a quadratic polynomial $Q = ax^2 + bx + c$.

Then, taking the discriminat D of Q, we get: $D = b^2 - 4ac$.

So D of P is as follows:

$$(\tfrac{112}{3})^2 - 4 \cdot 49 \cdot \tfrac{64}{9} = (\tfrac{112}{3})^2 - 2^2 \cdot 7^2 \cdot \tfrac{8^2}{3^2} = (\tfrac{112}{3})^2 - (2 \cdot 7 \cdot \tfrac{8}{3})^2 = (\tfrac{112}{3})^2 - (\tfrac{112}{3})^2 = 0.$$

Thus, P can be put in a complete square, and can be factorized the way below:

$$P = 49r^2 + \tfrac{112}{3}r + \tfrac{64}{9} = (7r)^2 + 2 \cdot \tfrac{8}{3} \cdot 7r + (\tfrac{8}{3})^2 = (7r + \tfrac{8}{3})^2.$$

And we can put it this way, too:

$$P = 49r^2 + \tfrac{112}{3}r + \tfrac{64}{9} = 49r^2 + \tfrac{56}{3}r + \tfrac{56}{3}r + \tfrac{64}{9} = 7r(7r + \tfrac{8}{3}) + \tfrac{8}{3}(7r + \tfrac{8}{3}) = (7r + \tfrac{8}{3})^2.$$

In short:

$$P = 49r^2 + \tfrac{112}{3}r + \tfrac{64}{9} = 49r^2 + \tfrac{56}{3}r + \tfrac{56}{3}r + \tfrac{64}{9} = 7r(7r + \tfrac{8}{3}) + \tfrac{8}{3}(7r + \tfrac{8}{3}) = (7r + \tfrac{8}{3})^2.$$

Examples 8

Factorize the polynomials below.

0. $a^2 + 10a + 9$

1. $\frac{4}{25}h^2 - \frac{17}{10}h + \frac{289}{64}$

2. $48q^2 - 168q + 147$

3. $\frac{2}{3}p^2 - 6p + \frac{27}{2}$

Suggestions or Solutions
To the **Problem** in the Example **0**

We have: $a^2 + 10a + 9$.

If a polynomial is a complete square, the factorization of it is easy.
That's because putting it into a complete square, we get the factorization of it.

Suppose now, $P = a^2 + 10a + 9$.   How then, can we see if P is a complete square?

Assuming for instance, $Q = ax^2 + bx + c$, we can say that the quadratic polynomial Q can be put in a complete square if the discriminant D of Q is 0, and $D = b^2 - 4ac$.

So getting the discriminant D of P, we get: $D = 10^2 - 4 \cdot 1 \cdot 9 = 100 - 36 \neq 0$.
So P cannot be put in a complete square.   What then?

We have a factorization identity, $x^2 + (u + v)x + uv = (x + u)(x + v)$, which has the same structure as the one the polynomial P has.   So?

So assuming the polynomial P is: $a^2 + (u + v)a + uv$, we can expect that P can be factorized to $(a + u)(a + v)$, where u and v are integers, $uv = 10$, and $u + v = 9$.

And if P is factorable within the scope of integers, using the divisors of 10, we can find by trial and error, a pair of integers that can be multiplied to be 10, and add up to 9.

We have: $10 = 1 + 9$, and $9 = 1 \cdot 9$.   So we can see that u and v are 1 and 9 respectively.

That is, $u = 1$ and $v = 9$, or $u = 9$ and $v = 1$.   Thus, we get:
$P = a^2 + 10a + 9 = a^2 + (1 + 9)a + 1 \cdot 9 = a^2 + a + 9a + 9$
$= a(a + 1) + 9(a + 1) = (a + 1)(a + 9)$.

In short:
$a^2 + 10a + 9 = a^2 + (1 + 9)a + 1 \cdot 9 = a^2 + a + 9a + 9 = a(a + 1) + 9(a + 1)$
$= (a + 1)(a + 9)$.

Suggestions or Solutions
To the **Problem** in the Example **1**

We have: $\frac{4}{25}h^2 - \frac{17}{10}h + \frac{289}{64}$.

Let's set first, $p = \frac{4}{25}h^2 - \frac{17}{10}h + \frac{289}{64}$.

The polynomial **P** looks quite messy, doesn't it?

However, the coefficient of h^2 is: $(\frac{2}{5})^2$, and the denominator of $\frac{289}{64}$ is 8^2.   So?

So we can reasonably expect that **P** can be put in a complete square.

So we may want to first, get the discriminant of **P**, and then, see if it is 0.

The discriminant is as follows:

$$(-\tfrac{17}{10})^2 - 4 \cdot \tfrac{4}{25} \cdot \tfrac{289}{64} = (\tfrac{17}{10})^2 - 4 \cdot \tfrac{4}{25} \cdot \tfrac{17^2}{8 \cdot 8} = (\tfrac{17}{10})^2 - \tfrac{1}{25} \cdot \tfrac{17^2}{2 \cdot 2} = (\tfrac{17}{10})^2 - \tfrac{17^2}{100} = 0.$$

So the factorization can go the way below:

$$p = \tfrac{4}{25}h^2 - \tfrac{17}{10}h + \tfrac{289}{64} = (\tfrac{2}{5}h)^2 - 2 \cdot \tfrac{17}{20}h + (\tfrac{17}{8})^2 = (\tfrac{2}{5}h)^2 - 2 \cdot \tfrac{2 \cdot 17}{2 \cdot 20}h + (\tfrac{17}{8})^2$$

$$= (\tfrac{2}{5}h)^2 - 2 \cdot \tfrac{2}{5} \cdot \tfrac{17}{8}h + (\tfrac{17}{8})^2 = (\tfrac{2}{5}h - \tfrac{17}{8})^2.$$

We can put it this way, too:

$$p = \tfrac{4}{25}h^2 - \tfrac{17}{10}h + \tfrac{289}{64} = (\tfrac{2}{5}h)^2 - \tfrac{17}{20}h - \tfrac{17}{20}h + (\tfrac{17}{8})^2 = \tfrac{2}{5}h(\tfrac{2}{5}h - \tfrac{17}{8}h) + \tfrac{17}{8}(-\tfrac{8}{20}h + \tfrac{17}{8})$$

$$= \tfrac{2}{5}h(\tfrac{2}{5}h - \tfrac{17}{8}h) - \tfrac{17}{8}(\tfrac{8}{20}h - \tfrac{17}{8}) = \tfrac{2}{5}h(\tfrac{2}{5}h - \tfrac{17}{8}h) - \tfrac{17}{8}(\tfrac{2}{5}h - \tfrac{17}{8}) = (\tfrac{2}{5}h - \tfrac{17}{8})^2.$$

In short:

$$\tfrac{4}{25}h^2 - \tfrac{17}{10}h + \tfrac{289}{64} = (\tfrac{2}{5}h)^2 - 2 \cdot \tfrac{17}{20}h + (\tfrac{17}{8})^2 = (\tfrac{2}{5}h)^2 - 2 \cdot \tfrac{2 \cdot 17}{2 \cdot 20}h + (\tfrac{17}{8})^2$$

$$= (\tfrac{2}{5}h)^2 - 2 \cdot \tfrac{2}{5} \cdot \tfrac{17}{8}h + (\tfrac{17}{8})^2 = (\tfrac{2}{5}h - \tfrac{17}{8})^2.$$

Suggestions or Solutions
To the **Problem** in the Example **2**

Suppose $P = 48q^2 - 168q + 147$.

Norammly, factorizing a polynomial, we begin with a divisor common to all the terms in the polynomial. Then, the divisor is called a common divisor, and is a factor of the polynomial.

And checking to see if an integer divides another integer, we can use a tool as below:

- If 3 divides the sum of all the digits in an integer, 3 divides the integer, too.

Now, in the polynomial P, the coefficients are 48, and -168, and the numeric term is 147.

So we may want to begin with a divisor common to all those integers.
Then, such a divisor can divide P, of course.
We can readily see that 2 cannot be a common factor since 147 is odd.
So let's see now, if 3 can do it.

For 48, we get: $4 + 8 = 12 = 3 \cdot 4$.

For 168, we get: $1 + 6 + 8 = 15 = 3 \cdot 5$.

For 147, we get: $1 + 4 + 7 = 12 = 3 \cdot 4$.

Thus, we can take 3 for the common factor, and we can reduce the polynomial to such a product form as follows: $P = 48q^2 - 168q + 147 = 3(16q^2 - 56q + 49)$.

Next, we can quckly notice that $16 = 4^2$, -56 is even, and $49 = 7^2$.
Thus, we can expect that the smaller polynomial, which is inside the parentheses, can be put in a complete square.

So keeping that in mind, let's factorize the polynomial.
Then, we get: $16q^2 - 56q + 49 = (4q)^2 - 2 \cdot 7 \cdot 4q + 7^2 = (4q - 7)^2$.
Thus, P gets factorized to $3(4q - 7)^2$.

In short:

$8q^2 - 168q + 147 = 3(16q^2 - 56q + 49) = (4q)^2 - 2 \cdot 7 \cdot 4q + 7^2 = (4q - 7)^2$.

Suggestions or Solutions
To the **Problem** in the Example **3**

We have: $\frac{2}{3}p^2 - 6p + \frac{27}{2}$.

Factorizing it, we can get it the way below:

$$\frac{2}{3}p^2 - 6p + \frac{27}{2} = \frac{1}{6} \cdot 6(\frac{2}{3}p^2 - 6p + \frac{27}{2}) = \frac{1}{6}(4p^2 - 36p + 81).$$

Meanwhile, $4p^2 - 36p + 81 = 4p^2 - 2 \cdot 9p - 2 \cdot 9p + 81 = 2p(2p-9) - 9(2p-9) = (2p-9)^2$.

Therefore, $\frac{2}{3}p^2 - 6p + \frac{27}{2} = \frac{1}{6}(2p-9)^2$.

If not quite sure of the idea behind the processes above, follow the steps below:

Running into a quadratic polynomial where coefficients or the constant term are not integers, we tend to think such a polynomial cannot be put in a complete square.

It is often the case though, such a polynomial can be put in a complete square.

We have already taken care of such polynomials earlier in this set of examples. In such a polynomial, the coefficient of the quadratic term is a number squared as $\frac{4}{25}$, and so is the numeric term. A quadratic term is a term where a variable is squared as $2x^2$.

Now, setting $Q = \frac{2}{3}p^2 - 6p + \frac{27}{2}$, and examining the coefficients and the numeric term in Q, we can see that they are not numbers squared, so it seems that Q cannot be put in a complete square.

Nevertheless, Q can still be put in a complete square.

So since Q is quadratic, we may want to check to see if its discriminant is 0. If so, it can be put in a complete square.   What is the discriminant though?

Assuming for instance, D is the discriminant of $ax^2 + bx + c$, we get: $D = b^2 - 4ac$.

And if $D = 0$, the polynomial $ax^2 + bx + c$ can be put in a complete square. How come?

The polynomial $ax^2 + bx + c$ is factorized to $a(x + \frac{b+\sqrt{D}}{2a})(x + \frac{b-\sqrt{D}}{2a})$, where $D = b^2 - 4ac$.

So if $D = b^2 - 4ac = 0$, we get: $ax^2 + bx + c = a(x + \frac{b}{2a})(x + \frac{b}{2a}) = a(x + \frac{b}{2a})^2$.

And thus, we may want to check first, to see if the discriminant D of Q is 0.

Then, D of Q is: $(-6)^2 - 4 \cdot \frac{2}{3} \cdot \frac{27}{2} = 36 - 36 = 0$, so Q can be put in a complete square.

What then, is the next?

Next, we have: $ax^2 + bx + c = a(x + \frac{b}{2a})^2$.

And in the case of Q, we have: $a = \frac{2}{3}$, and $b = -6$, so we get: $\frac{b}{2a} = \frac{-6}{2 \cdot \frac{2}{3}} = -\frac{18}{4} = -\frac{9}{2}$.

And thus, Q can be factorized to $\frac{2}{3}(p - \frac{9}{2})^2$.

Running into a polynomial where the coefficients and numeric term are messy as in the case of the polynomial Q, we can begin with simplifying such a polynomial.

We can take this approach when the coefficients and numeric term are rational numbers.

Suppose for instance, we want to factorize $U = \frac{1}{6}x^2 - \frac{2.3}{12}x - \frac{0.7}{4}$.

Then first, multiplying U by 10, we get: $10U = \frac{10}{6}x^2 - \frac{23}{12}x - \frac{7}{4}$.

Next, multiplying $10U$ by 12, which is the common denominator of $\frac{1}{6}$, $\frac{1}{12}$, and $\frac{1}{4}$, we get: $120U = 20x^2 - 23x - 21$, which is a polynomial without fractional numbers.

So in sum, multiplying and dividing U by 120, we get:

$$U = \tfrac{1}{6}x^2 - \tfrac{2.3}{12}x - \tfrac{0.7}{4} = 120 \cdot \tfrac{1}{120}(\tfrac{1}{6}x^2 - \tfrac{2.3}{12}x - \tfrac{0.7}{4}) = \tfrac{1}{120}(20x^2 - 23x - 21).$$

Next, for instance, setting: $V = 20x^2 - 23x - 21$, we can factorize V easily, and factorizing it, we get: $V = (5x + 3)(4x - 7)$.   What then, is the next?

We want to scale V back to the original one, which is U.

That is, we want to divide $V = (5x + 3)(4x - 7)$ by 120, and get U back.

So we get: $U = \tfrac{1}{120}(5x + 3)(4x - 7).$

How can we though, factorize such a polynomials as $20x^2 - 23x - 21$?

We will cover such a polynomial factorization in the **Examples B**.

Now in this problem, we are factorizing $Q = \tfrac{2}{3}p^2 - 6p + \tfrac{27}{2}$.

So taking the common denominator of $\tfrac{1}{3}$ and $\tfrac{1}{2}$, and multiplying and dividing Q by it, which is $\tfrac{1}{6}$, we get: $Q = \tfrac{2}{3}p^2 - 6p + \tfrac{27}{2} = \tfrac{1}{6} \cdot 6(\tfrac{2}{3}p^2 - 6p + \tfrac{27}{2}) = \tfrac{1}{6}(4p^2 - 36p + 81).$

So next, we want to factorize $4p^2 - 36p + 81$.

Examining the coefficients and the numeric term, we can see that $4 = 2^2$, $81 = 9^2$, and $36 = 2 \cdot 2 \cdot 9$, so we can expect that the polynomial above can be put in a complete square.

Actually getting the discriminant D, we get: $D = (-36)^2 - 4 \cdot 4 \cdot 81 = (4 \cdot 9)^2 - 4 \cdot 4 \cdot 9 \cdot 9 = 0.$

So it is the case. And thus, factorizing $4p^2 - 36p + 81$, we get:

$$4p^2 - 36p + 81 = 4p^2 - 4 \cdot 9p + 81 = 4p^2 - 2 \cdot 9p - 2 \cdot 9p + 81 = 2p(2p - 9) - 2 \cdot 9p + 81$$

$$= 2p(2p - 9) - 9(2p - 9) = (2p - 9)(2p - 9) = (2p - 9)^2.$$

So we get:

$$Q = \frac{2}{3}p^2 - 6p + \frac{27}{2} = \frac{1}{6}(4p^2 - 36p + 81) = \frac{1}{6}(2p - 9)^2 = \frac{1}{6}\{2(p - \frac{9}{2})\}^2 = \frac{4}{6}(p - \frac{9}{2})^2 = \frac{2}{3}(p - \frac{9}{2})^2.$$

In short:

$$\frac{2}{3}p^2 - 6p + \frac{27}{2} = \frac{1}{6} \cdot 6(\frac{2}{3}p^2 - 6p + \frac{27}{2}) = \frac{1}{6}(4p^2 - 36p + 81).$$

Meanwhile, $4p^2 - 36p + 81 = 4p^2 - 2 \cdot 9p - 2 \cdot 9p + 81 = 2p(2p - 9) - 9(2p - 9) = (2p - 9)^2.$

Therefore, $\frac{2}{3}p^2 - 6p + \frac{27}{2} = \frac{1}{6}(2p - 9)^2.$

Examples 9

Factorize the polynomials below.

0. $v^2 + 15v + 36$

1. $w^2 - 14w + 48$

2. $3a^2 + 12a - 135$

3. $t^2 + 5t + 5$, and $t^2 + 3t + 5$.

Suggestions or Solutions
To the **Problem** in the Example **0**

We have: $v^2 + 15v + 36$.

Let' set first, $P = v^2 + 15v + 36$.
Then, since P is quadratic, we may want to begin with checking to see if P can be put in a complete square. That is, we get the discriminant, and then, check to see if it is 0.

The discriminant of a polynomial $av^2 + bv + c$ where a, b, and c are constant is $b^2 - 4ac$, and thus, that of P is: $15^2 - 4 \cdot 1 \cdot 36$, which is not 0, because $4 \cdot 1 \cdot 36$ is even, but 15^2 is odd, since an odd number squared is odd. So P cannot be put in a complete square.

So let's factorize P in the usual manner.

To begin with, P is similar to $x^2 + (a + b)x + ab$, which is factorized to $(x + a)(x + b)$.

Thus, we can expect that P gets factorized to $(v + s)(v + t)$, where s and t are integers.

In other words, assuming that s and t are integers, we get to solve a system where $st = 36$ and $s + t = 15$. If any, we can find two integers adding up to 15 and getting multiplied to be 36 using divisors of 36 via trial and error.

If however, the values satisfying the system above cannot be integers, that is, s and t cannot be integers, we need to consult the quadratic formula, which is the solution to the quadratic equation $ax^2 + bx + c = 0$, and the quadratic formula is as follows:

$$x = \frac{-b \pm \sqrt{b^2 - 4ac}}{2a}, \text{ which is } x = \frac{-k \pm \sqrt{k^2 - ac}}{a} \text{ if } b = 2k.$$

We can put it the way below, too:

$$x = \frac{-b \pm \sqrt{D}}{2a} \text{ where } D = b^2 - 4ac, \text{ or if } b = 2k, x = \frac{-k \pm \sqrt{d}}{a} \text{ where } d = k^2 - ac.$$

Now, finding the pair of integers satisfying the system where $st = 36$, and $s + t = 15$, we want to begin with divisors of 36, and then, check to see if the sum of the divisors is 15.

We have: $36 = 12 \cdot 3$, and $12 + 3 = 15$.   So P can be factorized to $(v + 12)(v + 3)$.

In short:

$$v^2 + 15v + 36 = v^2 + 12v + 3v + 36 = v(v + 12) + 3(v + 12) = (v + 12)(v + 3).$$

Suggestions or Solutions
To the **Problem** in the Example **1**

We have: $w^2 - 14w + 48$.

Factorizing it, we can get it the way below:

$$w^2 - 14w + 48 = w^2 - 8w - 6w + 48 = w(w - 8) - 6(w - 8) = (w - 8)(w - 6).$$

If not quite sure of the idea behind the processes above, follow the steps below:

Let's set first, $P = w^2 - 14w + 48$.

The polynomial P is quadratic, so we may want to begin with checking to see if P can be put in a complete square. So we want to see if the discriminant is 0. The discriminant is:

$$(-14)^2 - 4 \cdot 1 \cdot 48 = 2 \cdot 7 \cdot 2 \cdot 7 - 4 \cdot 48 = 4 \cdot 7 \cdot 7 - 4 \cdot 48 = 4 \cdot 49 - 4 \cdot 48 = 4(49 - 48) = 4 \neq 0.$$

Thus, P cannot be put in a complete square.

So let's factorize P in the usual manner.

To begin with, P is similar to $x^2 - (a + b)x + ab$, which is factorized to $(x - a)(x - b)$.

However, P can be said to be similar to $x^2 + (a + b)x + ab$, too, which gets factorized to $(x + a)(x + b)$.

That is because changing a and b with $-a$ and $-b$ respectively in $(x - a)(x - b)$, we get:

$$(x - a)(x - b) = \{x - (-a)\}\{x - (-b)\} = (x + a)(x + b).$$

So $x^2 - (a + b)x + ab$ is just about the same as $x^2 + (a + b)x + ab$.

Thus, we can expect that P gets factorized to a polynomial in a form of $(w + s)(w + t)$, where s and t are integers.

In other words, assuming that s and t are integers, we get to solve a system where $st = 48$ and $s + t = -14$.

And if any, we can find two integers adding up to -14 and getting multiplied to be 48 using divisors of 48 via trial and error.

If however, no integers satisfy the system above, that is, s and t cannot be integers, we want to consult the quadratic formula.

We have: $48 = 6{\cdot}8$, but $6 + 8 \neq -14$, yet have this, too: $48 = (-6){\cdot}(-8)$, and $-6 + (-8) = -14$.

So the factorization of P can proceed the way below:

$$P = w^2 - 14w + 48 = w^2 - 8w - 6w + 48 = w(w - 8) - (6w - 48) = w(w - 8) - 6(w - 8)$$
$$= (w - 8)(w - 6).$$

In short:

$$w^2 - 14w + 48 = w^2 - 8w - 6w + 48 = w(w - 8) - 6(w - 8) = (w - 8)(w - 6).$$

Suggestions or Solutions
To the **Problem** in the Example **2**

We have: $3a^2 + 12a - 135$.

Factorizing it, we can get it the way below:

$$3a^2 + 12a - 135 = 3(a^2 + 4a - 45) = 3(a^2 - 5a + 9a - 45) = 3\{a(a - 5) + 9(a - 5)\}$$

$$= 3(a - 5)(a + 9).$$

If not quite sure of the idea behind the processes above, follow the steps below:

Setting first, $P = 3a^2 + 12a - 135$, we can say that the polynomial P is quadratic, so if the discriminant is 0, P can be put in a complete square.

Examining however, the coefficients and the numeric term, we can notice that 3 is a divisor common to all the terms in the polynomial P.

So we may want to first, take out 3 from the polynomial.

How do we know though, if 3 divides 135 without actually doing the divisions?

We can see, of course, if 3 divides 135 by doing the actual division.

We can also use though, such a tool as follows:

- If the sum of all the digits in an integer is a multiple of 3, 3 divides the integer.

The sum of all the digits in 135 is: $1 + 3 + 5 = 9$, which is a multiple of 3, so 3 divides it.

Thus, we get: $P = 3a^2 + 12a - 135 = 3(a^2 + 4a - 45)$.

Suppose now, Q is: $a^2 + 4a - 45$, and let's see if Q can be put in a complete square.

The discriminant of Q is: $2^2 - 1 \cdot (-45)$, which is not 0, and therefore, Q cannot be put in a complete square.

So let's factorize Q in the usual manner.

The polynomial Q is similar to $x^2 + (a + b)x + ab$, which is factorized to $(x + a)(x + b)$.

Thus, we can expect that Q gets factorized to $(a + s)(a + t)$, where s and t are integers.

That is to say that assuming that s and t are integers, we get to solve a system of two equations where $st = -45$ and $s + t = 4$. If any, we can find two integers adding up to 4 and getting multiplied to be -45 using divisors of -45 via trial and error.

If however, no integers satisfy the system above, that is, s and t cannot be integers, we want to consult the quadratic formula.

Now, to begin with, we can get: $-45 = 5 \cdot (-9)$, but we get: $5 + (-9) \neq 4$.

And next, we get: $-45 = (-5) \cdot 9$, and $-5 + 9 = 4$.

So the factorization of P can proceed as follows:

$$P = 3a^2 + 12a - 135 = 3(a^2 + 4a - 45)$$

$$= 3(a^2 - 5a + 9a - 45) = 3\{a(a - 5) + 9(a - 5)\} = 3(a - 5)(a + 9).$$

Suggestions or Solutions
To the Problem in the Example 3

We have: $t^2 + 5t + 5$, and $t^2 + 3t + 5$.

To begin with, setting $t^2 + 5t + 5 = 0$, we get: $t = \frac{-5 \pm \sqrt{D}}{2}$, where $D = 5^2 - 4 \cdot 1 \cdot 5 = 5$.

Thus, $t^2 + 5t + 5 = (t - \frac{-5+\sqrt{5}}{2})(t - \frac{-5-\sqrt{5}}{2}) = (t + \frac{5-\sqrt{5}}{2})(t + \frac{5+\sqrt{5}}{2})$.

Next, the discriminant of $t^2 + 3t + 5$ is: $3^2 - 4 \cdot 1 \cdot 5 = 9 - 20 = -11$.

Thus, $t^2 + 3t + 5$ is not factorable (within the scope of real numbers).

If not quite sure of the idea behind the processes above, follow the steps below:

Let's set first, $P = t^2 + 5t + 5$, and $Q = t^2 + 3t + 5$.

The polynomials P and Q are quadratic, so we may want to begin with checking to see if they can be put in complete squares. So we want to see if the discriminants of both are 0.

The discriminant of P is: $5^2 - 4 \cdot 1 \cdot 5 = 5(5 - 4) = 5$, which is not 0.

The discriminant of Q is: $3^2 - 4 \cdot 1 \cdot 5 = 9 - 20 = -11$, which is not 0 either.

So we can put neither of P and Q in a complete square, and therefore, we may want to factorize both of those in the usual manner.

First, both are similar to $x^2 + (a + b)x + ab$, which we can factorize to $(x + a)(x + b)$.

So next, we can expect each polynomial can get factorized to $(t + u)(t + v)$, where u and v are integers, since the coefficients and the numeric term are integers.

Thus, next, beginning with P, we can set: $uv = 5$, and $u + v = 5$, which make a system of equations for u and v.   If u and v cannot be integers satisfying the system of equations above, we need to use the quadratic formula. In fact, there is no pair of integers satisfying the system.   How come?

Taking advantage of the discriminant, usually designated by D, we can readily check to see if it is the case.   Where is the discriminant D from, though?

It is from the quadratic formula, isn't it?

So let's take a closer look at the formula, which is: $x = \frac{-b \pm \sqrt{D}}{2a}$, where $D = b^2 - 4ac$.

The formula is for a quadratic equation $ax^2 + bx + c = 0$, where a, b, and c are constant.

However, the two polynomials P and Q are in a form of $x^2 + mx + n$, where m and n are integers.

So applying the formula to a quadratic equation $x^2 + mx + n = 0$, we can set first, $a = 1$, $b = m$, and $c = n$, and thus next, we can get:

$$x = \frac{-b \pm \sqrt{D}}{2a} = \frac{-m \pm \sqrt{D}}{2 \cdot 1} = \frac{-m \pm \sqrt{D}}{2}, \text{ where } D = b^2 - 4ac = m^2 - 4 \cdot 1 \cdot n = m^2 - 4n.$$

Thus, we get: $x = \frac{-m \pm \sqrt{D}}{2}$, where $D = m^2 - 4n$.

So we get: $x^2 + mx + n = (x - \frac{-m+\sqrt{D}}{2})(x - \frac{-m-\sqrt{D}}{2}) = (x + \frac{m-\sqrt{D}}{2})(x + \frac{m+\sqrt{D}}{2}) = 0$.

Thus, we get: $x^2 + mx + n = (x + \frac{m-\sqrt{D}}{2})(x + \frac{m+\sqrt{D}}{2}) = 0$, where $D = m^2 - 4n$.

Now, our expectation is that each of P and Q can get factorized to a form of $(t + u)(t + v)$, where u and v are integers.

The two polynomials P and Q are in a form of $x^2 + mx + n$, where m and n are integers.

We have: $x^2 + mx + n = (x + \frac{m-\sqrt{D}}{2})(x + \frac{m+\sqrt{D}}{2}) = 0$, where $D = m^2 - 4n$.

So replacing x with t, we get: $t^2 + mt + n = (t + \frac{m-\sqrt{D}}{2})(t + \frac{m+\sqrt{D}}{2}) = 0$, where $D = m^2 - 4n$.

Thus, we can see that $u = \frac{m-\sqrt{D}}{2}$, and $v = \frac{m+\sqrt{D}}{2}$, where $D = m^2 - 4n$.

Now, we know m and n are integers.
So the discriminant D is an integer because m and n are integers.

That is because we get an integer doing a multiplication with integers and doing a subtraction with integers as well as doing an addition with them. So a multiplication, a subtraction, or an addition with integers produces an integer.

Now, if the values of u and v are integers, D has to be an integer squared as 3^2 since D is inside the square root sign. So let's have a look at what D the polynomial P has.

The discriminant D of P is: $5^2 - 4 \cdot 1 \cdot 5 = 5(5 - 4) = 5$, which is not an integer squared.

So the values of u and v cannot be integers.
Thus, there is no pair of integers satisfying the system where $uv = 5$, and $u + v = 5$.
Therefore, factorizing P, we need to use the quadratic formula.

Then, setting $P = t^2 + 5t + 5 = 0$, we get: $t = \frac{-5 \pm \sqrt{D}}{2}$, where $D = 5^2 - 4 \cdot 1 \cdot 5 = 5$.

So we get: $P = t^2 + 5t + 5 = (t - \frac{-5+\sqrt{5}}{2})(t - \frac{-5-\sqrt{5}}{2}) = (t + \frac{5-\sqrt{5}}{2})(t + \frac{5+\sqrt{5}}{2})$.

Thus, we get: $P = t^2 + 5t + 5 = (t + \frac{5-\sqrt{5}}{2})(t + \frac{5+\sqrt{5}}{2})$.

So we can see that P can get factorized to $(t + \frac{5-\sqrt{5}}{2})(t + \frac{5+\sqrt{5}}{2})$.

- Let's next, move on to the polynomial $Q = t^2 + 3t + 5$.

Either checking to see if Q can be put in a complete square, or checking to see if Q can get factorized to $(t + u)(t + v)$ where u and v are integers, we can take advantage of the discriminant D.

The discriminant D of Q is: $3^2 - 4 \cdot 1 \cdot 5 = 9 - 20 = -11$, which is neither 0 nor an integer squared.

So Q cannot be put in a complete square, and cannot be factorized to $(t + u)(t + v)$ where u and v are integers.   That's not it, though.

The polynomial Q is not factorable.   That is to say that Q has no divisor.
In other words, Q itself is a prime polynomial.   How come?

We know the solution to $ax^2 + bx + c = 0$ is $x = \frac{-b \pm \sqrt{D}}{2a}$, where D is the discriminant.

Now, if the discriminant D is negative, D cannot be inside a square root sign.

In real number space, no negative number can be inside a square root sign.

In other words, the equation $Q = t^2 + 3t + 5 = 0$ can have no solution.

Therefore, Q is not factorable.
More precisely, Q cannot be factorable in real number space, yet it can be factorable in complex number space, because a square root of a negative number is an imaginary number, which can exist not in real number space but in complex number space.

So we have:

- If a quadratic polynomial is factorable, the discriminant is greater than or equal to 0.

- If a polynomial $x^2 + mx + n$ where m and n are integers is factorized to $(x + u)(x + v)$ where u and v are integers, the discriminant is an integer squared as 3^2.

What is the real number space though?

It's a space where all real numbers exist, and real numbers only can exist.
So it is a set of all real numbers.

Real numbers include rational numbers and irrational numbers as $\pm\sqrt{2}, \pm\pi$, where π is the circular ratio, which is $3.141592\ldots$, and rational numbers include integers and fractional numbers as 0.2, -0.3, 3.5, $\frac{1}{3}$, $\frac{1}{4}$, $\frac{-9}{2}$, etc.

However, the complex number space is a space where any number can exist, and thus, is the entire number space.

In short:

To begin with, setting $t^2 + 5t + 5 = 0$, we get: $t = \frac{-5 \pm \sqrt{D}}{2}$, where $D = 5^2 - 4 \cdot 1 \cdot 5 = 5$.

Thus, $t^2 + 5t + 5 = (t - \frac{-5+\sqrt{5}}{2})(t - \frac{-5-\sqrt{5}}{2}) = (t + \frac{5-\sqrt{5}}{2})(t + \frac{5+\sqrt{5}}{2})$.

Next, the discriminant of $t^2 + 3t + 5$ is: $3^2 - 4 \cdot 1 \cdot 5 = 9 - 20 = -11$.

Thus, $t^2 + 3t + 5$ is not factorable (within the scope of real numbers).

Examples A

Find the values of *a* and *b* in each of the cases below.

0. $a^2 + b^2 = 108$, and $ab = 18$.

1. $a - b = 10$, and $a^2 + b^2 = 58$.

2. $a + b = 12$, and $a^2 - b^2 = 168$.

Suggestions or Solutions

Find the values of a and b for which $a^2 + b^2 = 85$, and $ab = 42$.

$(a + b)^2 = a^2 + 2ab + b^2 = a^2 + b^2 + 2ab = 85 + 2 \cdot 42 = 85 + 84 = 169 = 13^2$.
So we get: $a + b = \pm 13$, and $ab = 42$.

$a + b = 13 \Rightarrow a = 13 - b \Rightarrow ab = (13 - b)b = 42 \Rightarrow 13b - b^2 = 42 \Rightarrow b^2 - 13b + 42 = 0$.
$b^2 - 13b + 42 = (b - 6)(b - 7) = 0 \Rightarrow b = 6$ or 7.

$b = 6 \Rightarrow a = 13 - b = 13 - 6 = 7$. $b = 7 \Rightarrow a = 13 - b = 13 - 7 = 6$.
Thus, we get: $(a, b) = (6, 7)$ or $(7, 6)$.

$a + b = -13 \Rightarrow a = -13 - b \Rightarrow ab = (-13 - b)b = -13b - b^2 = 42 \Rightarrow b^2 + 13b + 42 = 0$.
$b^2 + 13b + 42 = (b + 6)(b + 7) = 0 \Rightarrow b = -6$ or -7.

$b = -6 \Rightarrow a = -13 - b = -13 + 6 = -7$. $b = -7 \Rightarrow a = -13 - b = -13 + 7 = -6$.
Thus, we get: $(a, b) = (-6, -7)$ or $(-7, -6)$.

Therefore, $(a, b) = (6, 7)$, $(7, 6)$, $(-6, -7)$, or $(-7, -6)$.

If not quite sure of the idea behind the processes above, follow the steps below:

We are given a system of equations, and the equations are: $a^2 + b^2 = 85$, and $ab = 42$.
And the solution can be made in such an ordinary way as follows:

Assuming $b \neq 0$, and expressing a in terms of b using $ab = 42$, we can get: $a = \frac{42}{b}$.

Then, putting it into $a^2 + b^2 = 85$, we get: $\left(\frac{42}{b}\right)^2 + b^2 = 85$.

And then, we can solve for b, the equation above.

Approaching however, the solution the way above, we get to do calculations quite messy and involved. In fact, the last equation for b above is of degree 4, which is pretty high. So we may want to take some other way.

Studying polynomial factorizations, we want to pay attention to not only the results but the structures of polynomials being factorized, together with factorization processes, too.

Having done a substantial amount of practice on polynomial factorizations paying much attention to such structures and processes, we can notice that the two equations given are elements of a particular polynomial. What polynomial then, is it?

It is the polynomial that is factorized to $(a + b)^2$, and thus, is: $a^2 + 2ab + b^2$. What then, are the elements stated above?

They are: $a^2 + b^2$ and ab, of which we have the values. So we get:

$$(a + b)^2 = a^2 + 2ab + b^2 = a^2 + b^2 + 2ab = 85 + 2 \cdot 42 = 85 + 84 = 169 = 13^2.$$

Thus, we can see that $(a + b) = 13^2$, and therefore, we get: $a + b = \pm 13$.

So we now have a new system where $a + b = \pm 13$, and $ab = 42$, which looks much simpler than the given system where $a^2 + b^2 = 85$, and $ab = 42$.

In other words, we just have reduced the system where $a^2 + b^2 = 85$, and $ab = 42$ to the new system where $a + b = \pm 13$, and $ab = 42$.

And solving the new system, since we have: $a + b = \pm 13$, we want to consider two cases where: $a + b = 13$, and $a + b = -13$, together with $ab = 42$.

So we get two systems as follows.

One of the two has: $a + b = 13$, and $ab = 42$, and the other has: $a + b = -13$, and $ab = 42$.

Beginning with the system where $a + b = 13$, and $ab = 42$, we can get first:

$$a + b = 13 \Rightarrow a = 13 - b \Rightarrow ab = (13 - b)b = 42 \Rightarrow 13b - b^2 = 42 \Rightarrow b^2 - 13b + 42 = 0,$$

which is noting but a simple quadratic equation for b.   So?

We can get the solution by a factorization (or by the quadratic formula, of course).

So factorizing $b^2 - 13b + 42$, we can get the solution to $b^2 - 13b + 42 = 0$.

Then, we want to find two integers getting multiplied to be 42, and adding up to -13.

By trial and error, using divisors of 42, we can get: $42 = (-6) \cdot (-7)$, and $-6 + (-7) = -13$.

So we get: $b^2 - 13b + 42 = (b - 6)(b - 7) = 0 \Rightarrow b = 6$ or 7.

And thus, we now have: $a = 13 - b$, and $b = 6$ or 7.   So we get:

$b = 6 \Rightarrow a = 13 - b = 13 - 6 = 7$.
$b = 7 \Rightarrow a = 13 - b = 13 - 7 = 6$.

Therefore, we get: $(a, b) = (6, 7)$ or $(7, 6)$.

• Let's next, move on to the other system where $a + b = -13$, and $ab = 42$.

Then, we can get first:

$$a + b = -13 \Rightarrow a = -13 - b \Rightarrow ab = (-13 - b)b = -13b - b^2 = 42 \Rightarrow b^2 + 13b + 42 = 0,$$

which is no more than a simple quadratic equation for b.

So factorizing $b^2 + 13b + 42$, we can get the solution to $b^2 + 13b + 42 = 0$.
Then, we want to find divisors of 42, of which the sum is 13.

By trial and error, we can get: $42 = 6 \cdot 7$, and $6 + 7 = 13$.

So we get: $b^2 + 13b + 42 = (b + 6)(b + 7) = 0 \Rightarrow b = $ -6 or -7.

And thus, we now have: $a = $ -13 $ - b$, and $b = $ -6 or -7.   So we get:

$b = $ -6 $\Rightarrow a = $ -13 $ - b = $ -13 $ + 6 = $ -7.
$b = $ -7 $\Rightarrow a = $ -13 $ - b = $ -13 $ + 7 = $ -6.

Therefore, we get: $(a, b) = $ **(-6, -7)** or **(-7, -6)**.

And putting threads together now, we get: $(a, b) = $ **(6, 7)**, **(7, 6)**, **(-6, -7)**, or **(-7, -6)**.

So studying polynomial factorizations, we want to pay attention to not only results but the structures of polynomials being factorized, together with factorization processes, too.

In short:

$(a + b)^2 = a^2 + 2ab + b^2 = a^2 + b^2 + 2ab = 85 + 2 \cdot 42 = 85 + 84 = 169 = 13^2$.
So we get: $a + b = \pm 13$, and $ab = 42$.

$a + b = 13 \Rightarrow a = 13 - b \Rightarrow ab = (13 - b)b = 42 \Rightarrow 13b - b^2 = 42 \Rightarrow b^2 - 13b + 42 = 0$.
$b^2 - 13b + 42 = (b - 6)(b - 7) = 0 \Rightarrow b = $ 6 or 7.

$b = 6 \Rightarrow a = 13 - b = 13 - 6 = 7$. $b = 7 \Rightarrow a = 13 - b = 13 - 7 = 6$.

Thus, we get: $(a, b) = $ **(6, 7)** or **(7, 6)**.

$a + b = $ -13 $\Rightarrow a = $ -13 $ - b \Rightarrow ab = ($ -13 $ - b)b = $ -13$b - b^2 = 42 \Rightarrow b^2 + 13b + 42 = 0$.
$b^2 + 13b + 42 = (b + 6)(b + 7) = 0 \Rightarrow b = $ -6 or -7.

$b = $ -6 $\Rightarrow a = $ -13 $ - b = $ -13 $ + 6 = $ -7. $b = $ -7 $\Rightarrow a = $ -13 $ - b = $ -13 $ + 7 = $ -6.

Thus, we get: $(a, b) = $ **(-6, -7)** or **(-7, -6)**.

Therefore, $(a, b) = $ **(6, 7)**, **(7, 6)**, **(-6, -7)**, or **(-7, -6)**.

• And let's now talk about some more algebra.

Once we have found a pair of values for a and b in the system where $a^2 + b^2 = 85$, and $ab = 42$, we can quickly find the other three pairs reflecting the two facts below:

• First, we can notice that the equation $a^2 + b^2 = 85$ remains the same even if a and b get exchanged. So we can see that a gets values b can get, and vice versa. It doesn't necessarily mean though, that a and b are the same.

• Next, the equation $ab = 42$ tells us the product of a and b is positive, so a and b have the same sign. That is, a and b both are positive, or are negative at the same time.
So once we have found a pair of values for a and b, we can quickly find the other three pairs using the two facts above.

Suppose for instance, we have found: $(a, b) = (6, 7)$.

Then, we can immediately see that $(a, b) = (7, 6)$, $(-6, -7)$, or $(-7, -6)$, too.

That's because:

First, a gets values that b can get, and vice versa.

So a can be **7** since $b = 7$, and b can be **6** since $a = 6$.

Thus, we can get: $(a, b) = (7, 6)$, too.

Next, a and b can be negative, too, at the same time.

Thus, we can have: $(a, b) = (-6, -7)$ or $(-7, -6)$, too.

So if in a multiple-choice test, we get to choose the solution to a system of equations like the system where $a^2 + b^2 = 85$, and $ab = 42$, we can cut corners the way above.

Suggestions or Solutions
To the **Problem 1** in the Example **1**

Find the values of a and b for which $a^2 + b^2 = 58$, and $a - b = 10$.

$(a - b)^2 = a^2 - 2ab + b^2 = a^2 + b^2 - 2ab = 58 - 2ab = 10^2 \Rightarrow 29 - ab = 50 \Rightarrow ab = -21$.

So we get: $a - b = 10$, and $ab = -21$.

Thus, we get: $a - b = 10 \Rightarrow a = b + 10 \Rightarrow ab = (b + 10)b = -21 \Rightarrow b^2 + 10b + 21 = 0$.

So we get: $b^2 + 10b + 21 = (b + 3)(b + 7) = 0 \Rightarrow b = -3$ or -7.

$b = -3 \Rightarrow a = b + 10 = -3 + 10 = 7$.
$b = -7 \Rightarrow a = b + 10 = -7 + 10 = 3$.

Therefore, we get: $(a, b) = (7, -3)$ or $(3, -7)$.

If not quite sure of the idea behind the processes above, follow the steps below:

We are given a system of equations, which are: $a - b = 10$, and $a^2 + b^2 = 58$.

Unlike the system in the problem 0, the system above can just be solved in a usual manner without much of complexity in calculation.

• Expressing a in terms of b using $a - b = 10$, we get: $a = b + 10$, which we can put into $a^2 + b^2 = 58$, and then, we solve: $(b + 10)^2 + b^2 = 58$.   Solving it, we can factorize it or use the quadratic formula.

Let's for practice purpose though, approach the solution the way we get the solution to the previous system.

Examining the two equations in the system given here, we can notice that one of the two equations is an element of a particular polynomial, which is factorized to $(a - b)^2$. What then, is the particular polynomial?

It is: $a^2 - 2ab + b^2$. And we have: $a^2 + b^2 = 58$, which is the element stated above. What about ab, though?

We can find the value of ab using $a - b = 10$, together with $a^2 + b^2 = 58$.

And using the values of $a - b$ and $a^2 + b^2$, we get:

$(a - b)^2 = a^2 - 2ab + b^2 = a^2 + b^2 - 2ab \Rightarrow 10^2 = 58 - 2ab \Rightarrow 2ab = -42 \Rightarrow ab = -21$.

So we just have reduced the given system where $a - b = 10$, and $a^2 + b^2 = 58$ to a new system where $a - b = 10$, and $ab = -21$, which is much simpler than the system given.

Now, solving the new system above, we can get first:

$a - b = 10 \Rightarrow a = b + 10 \Rightarrow ab = (b + 10)b = -21 \Rightarrow b^2 + 10b + 21 = 0$, which is noting but a simple quadratic equation for b.

So factorizing $b^2 + 10b + 21$, we can get the solution to $b^2 + 10b + 21 = 0$.

First, finding divisors of 21, of which the sum is 10, we can get: $21 = 3 \cdot 7$, and $3 + 7 = 10$.

So we get: $b^2 + 10b + 21 = (b + 3)(b + 7) = 0 \Rightarrow b = -3$ or -7.

Thus next, since we have: $a = b + 10$, we get:

$b = -3 \Rightarrow a = b + 10 = -3 + 10 = 7$.
$b = -7 \Rightarrow a = b + 10 = -7 + 10 = 3$.

Therefore, we get: $(a, b) = (7, -3)$ or $(3, -7)$.

Suggestions or Solutions
To the **Problem 2** in the Example 1

Find the values of *a* and *b* for which $a^2 - b^2 = 168$, and $a + b = 12$.

$a + b = 12 \Rightarrow a = 12 - b \Rightarrow a^2 - b^2 = (12 - b)^2 - b^2 = 168$

$\Rightarrow (12 - b)^2 - b^2 = 12^2 - 24b + b^2 - b^2 = 12^2 - 24b = 168 \Rightarrow 24b = 144 - 168 \Rightarrow b = -1$

$\Rightarrow a + b = a - 1 = 12 \Rightarrow a = 13$.

Therefore, $(a, b) = (13, -1)$.

If not quite sure of the idea behind the processes above, follow the steps below:

We are given a system of equations where $a + b = 12$, and $a^2 - b^2 = 168$.
As in the case of the problem 1, we can just directly approach the solution to the system above by an ordinary method. And the ordinary method is as follows:

• Putting *a* in terms of *b* using $a - b = 12$, we get: $a = 12 - b$, which we can put into the other equation $a^2 - b^2 = 168$, and then, we solve: $(12 - b)^2 - b^2 = 168$. Solving it though, we don't even need to bother factorizing it or using the quadratic formula.

That's because the quadratic (b^2) terms will get canceled out.
Simplifying the final equation above, we get:

$(12 - b)^2 - b^2 = 12^2 - 24b + b^2 - b^2 = 12^2 - 24b = 168 \Rightarrow 24b = 144 - 168 \Rightarrow b = -1$.

So the final equation is not really quadratic, and is just a simple equation of degree 1.

Let's for practice purpose though, approach the solution the way we get the solution to the system in the Problem 0.

What polynomial factorization then, can we refer to?

We can refer to the factorization of $a^2 - b^2$.

And factorizing it, we get: $a^2 - b^2 = (a + b)(a - b)$, which is called a factorization identity. And in the identity, we can see two polynomials, which are in the system given.

What then, are the two?

The two are: $a + b$, and $a^2 - b^2$, of which we have the values.

What about the value of $a - b$, though?

We can find the value of $a - b$ using $a + b = 12$, along with $a^2 - b^2 = 168$.

And taking advantage of the values of $a + b$ and $a^2 - b^2$, we get:

$$a^2 - b^2 = (a + b)(a - b) \Rightarrow 168 = 12(a - b) \Rightarrow a - b = 14.$$

So we now have reduced the given system to a system where $a - b = 14$, and $a + b = 12$, which is a simple system of equations of degree 1, which can be readily solved.

To begin with, adding up the two equations in the system, we get:

$$(a - b) + (a + b) = 14 + 12 \Rightarrow 2a = 26 \Rightarrow a = 13.$$

Next, subtracting one equation from the other, we can get:

$$(a - b) - (a + b) = 14 - 12 \Rightarrow -2b = 2 \Rightarrow b = -1.$$

Therefore, we get: $(a, b) = (13, -2)$.

In short:

$a + b = 12 \Rightarrow a = 12 - b \Rightarrow a^2 - b^2 = (12 - b)^2 - b^2 = 168$

$\Rightarrow (12 - b)^2 - b^2 = 12^2 - 24b + b^2 - b^2 = 12^2 - 24b = 168 \Rightarrow 24b = 144 - 168 \Rightarrow b = -1$

$\Rightarrow a + b = a - 1 = 12 \Rightarrow a = 13.$

Therefore, $(a, b) = (13, -1).$

And we can get the same the way below, too:

$a^2 - b^2 = (a + b)(a - b) \Rightarrow 168 = 12(a - b) \Rightarrow a - b = 14.$

$(a - b) + (a + b) = 14 + 12 \Rightarrow 2a = 26 \Rightarrow a = 13.$

$(a - b) - (a + b) = 14 - 12 \Rightarrow -2b = 2 \Rightarrow b = -1.$

Therefore, $(a, b) = (13, -2).$